Springer Theses

Recognizing Outstanding Ph.D. Research

For further volumes:
http://www.springer.com/series/8790

Aims and Scope

The series "Springer Theses" brings together a selection of the very best Ph.D. theses from around the world and across the physical sciences. Nominated and endorsed by two recognized specialists, each published volume has been selected for its scientific excellence and the high impact of its contents for the pertinent field of research. For greater accessibility to non-specialists, the published versions include an extended introduction, as well as a foreword by the student's supervisor explaining the special relevance of the work for the field. As a whole, the series will provide a valuable resource both for newcomers to the research fields described, and for other scientists seeking detailed background information on special questions. Finally, it provides an accredited documentation of the valuable contributions made by today's younger generation of scientists.

Theses are accepted into the series by invited nomination only and must fulfill all of the following criteria

- They must be written in good English.
- The topic should fall within the confines of Chemistry, Physics, Earth Sciences, Engineering and related interdisciplinary fields such as Materials, Nanoscience, Chemical Engineering, Complex Systems and Biophysics.
- The work reported in the thesis must represent a significant scientific advance.
- If the thesis includes previously published material, permission to reproduce this must be gained from the respective copyright holder.
- They must have been examined and passed during the 12 months prior to nomination.
- Each thesis should include a foreword by the supervisor outlining the significance of its content.
- The theses should have a clearly defined structure including an introduction accessible to scientists not expert in that particular field.

Malte Oppermann

Resolving Strong Field Dynamics in Cation States of CO_2 via Optimised Molecular Alignment

Doctoral Thesis accepted by
Imperial College London, UK

Springer

Author
Dr. Malte Oppermann
Department of Physics
Imperial College London
South Kensington, London
UK

Supervisor
Prof. Jonathan Marangos
Department of Physics
Imperial College London
South Kensington, London
UK

ISSN 2190-5053 ISSN 2190-5061 (electronic)
ISBN 978-3-319-35886-4 ISBN 978-3-319-05338-7 (eBook)
DOI 10.1007/978-3-319-05338-7
Springer Cham Heidelberg New York Dordrecht London

Printed on acid-free paper

Springer is part of Springer Science+Business Media (www.springer.com)

You can crank up the voltage even more. If you dare.

Leszek Frasinski about the only available detector for the experiments.

London, 2011

Supervisor's Foreword

The behaviour of matter in an intense laser, with electric field strengths near or exceeding 1 V/Å, is a frontier topic in strong field physics. This high field regime challenges our quantum theoretical and computational methods as the situation is inherently many-bodied and non-perturbative. It is a topic that is fascinating as it may offer new light sources, such as those based on high harmonic generation (HHG) with attosecond temporal structure as discussed by Paul et al. [1] and many other authors. The resulting fragmentation in molecules may also offer new routes to strong field chemical control (see for example Levis et al. [2]) and biomolecular analysis by selective bond cleavage as suggested by Weinkauf et al. [3].

The current experimental approaches concentrate on small gas phase molecules to build up ideas as to how the strong field influences the motion of the electrons and nuclei. It is essential to obtain high quality quantitative experimental data to feed into the advances in establishing theoretically how to understand the dynamics of molecules exposed to strong laser fields. One picture that greatly assists us is the so-called *strong field approximation* as developed formally by Lewenstein et al. [4], which captures the fact that the laser field, rather than the molecular electronic potential, dominates the behaviour of the ionised electron. An important feature of the strong field limit is the existence of laser-driven electron recollision that can lead to HHG and non-sequential double ionisation (NSDI) of an atom or molecule. The extent to which the strong field picture and electron recollision-induced phenomena can be used to understand the results of experiments on simple molecules like CO_2 is the main subject of Malte Oppermann's thesis.

What Malte has achieved is to conduct a series of measurements that explore the dependence of fragmentation and double ionisation on the alignment of the CO_2 molecular axis with respect to the polarisation direction of the strong field. To do this he employed impulsive molecular alignment of CO_2 molecules with a carefully timed pre-pulse of weaker intensity from the same titanium sapphire 800 nm laser used to generate the high intensity pulse (a topic well described by

Stapelfeldt and Seideman [5]). The technique required careful optimisation of the sample cooling and the 800 nm laser beam parameters to implement a high degree of alignment which was achieved by the use of a buffer gas (Ar) that resulted in very high degrees of cooling and thus alignment [6]. A Wiley–McLaren ion time of flight spectrometer was used to measure the fragment yield. Malte utilised the peaks from the Ar buffer gas for in situ monitoring of intensity fluctuations and non-sequential double ionization in the molecule.

Malte's first measurements with 800 nm strong fields showed weak non-sequential double ionisation in CO_2. He identified the higher energy recollision obtained with a longer wavelength field would increase NSDI. He then used an intense short pulse at 1,350 nm to see very clear signatures of NSDI in CO_2. He was able to measure the NSDI yield as a function of laser ellipticity, to confirm the signal came predominately from electron recollision, and then record the yield as a function of molecular alignment. From simple geometrical considerations he then deduced that his measured results could only be explained if the cation intermediate state HOMO and HOMO-1 were involved in the process, a new result published in [7]. His work has gone much further and looked at other zero kinetic energy release fragmentation products ($CO^+ + O$ and $O^+ + CO$) and his results support the idea that laser-driven transitions in the cation state must be invoked to excite the molecular ion to HOMO-2 and HOMO-3 from where it can dissociate.

He has used a high degree of ingenuity to use a simple apparatus to make the first measurements on the angular dependence of NSDI and fragmentation channels in CO_2, which are likely to be of considerable benefit to the theory community trying to develop accurate approaches to molecular strong field processes. Moreover, he has written a clear thesis that succinctly reviews the state-of-the-art in strong field molecular science.

London, January 2014 Jonathan Marangos

References

1. P.M. Paul, E.S. Toma, P. Breger, G. Mullot, F. Aug, P. Balcou, H.G. Muller, P. Agostini, Observation of a train of attosecond pulses from high harmonic generation. Science **292**, 1689–1692 (2001)
2. R.J. Levis, G.M. Menkir, H. Rabitz, Selective bond dissociation and rearrangement with optimally tailored, strong-field laser pulses. Science **292**, 709–713 (2001)
3. R. Weinkauf, E.W. Schlag, T.J. Martinez, R.D. Levine, Nonstationary electronic states and site-selective reactivity. The J. Phys. Chem. A **101**(42), 7702–7710 (1997)

4. M. Lewenstein, P. Balcou, M.Y. Ivanov, A. L'Huillier, P.B. Corkum, Theory of high-harmonic generation by low-frequency laser fields. Phys. Rev. A **49**, 2117 (1994)
5. H. Stapelfeldt, T. Seideman, Colloquium: aligning molecules with strong laser pulses. Rev. Mod. Phys. **75**, 543 (2003)
6. M. Oppermann, S.J. Weber, J.P. Marangos, Characterising and optimising impulsive molecular alignment in mixed gas samples. Phys. Chem. Chem. Phys. **14**(27), 9785–9791 (2012)
7. M. Oppermann, S.J. Weber, L.J. Frasinski, M.Y. Ivanov, J.P. Marangos, Multichannel contributions in the nonsequential double ionization of CO_2. Phys. Rev. A **88**, 043432 (2013)

Abstract

In this thesis, the role of the molecular structure in strong field induced processes in CO_2 is resolved by the use of optimised impulsive molecular alignment. Two processes were investigated: recollision induced ionisation and the dissociation of the molecular ion CO_2^+. Both processes are driven by the initial tunneling ionisation of CO_2. Through the use of molecular alignment, the orbital symmetries of the involved molecular ionic states were resolved, revealing the internal molecular dynamics in the form of ionisation and excitation pathways.

A novel data analysis procedure was developed to extract the alignment distribution and rotational temperature of the impulsively aligned molecular ensemble. This facilitated the optimisation of molecular alignment in mixed gas samples and was demonstrated for N_2, O_2 and CO_2 seeded in Ar.

The recollision induced or non-sequential double ionisation (NSDI) of CO_2 was then studied in the molecular frame. The process was fully characterised by measuring the shape of the recolliding electron wavepacket and the angularly resolved inelastic electron ion recollision cross-section. The results reveal the contribution from both the ionic ground and first excited state of CO_2^+ to the NSDI mechanism.

This study was extended to the strong field induced dissociation of impulsively aligned CO_2^+. It was found that dissociation is driven by a parallel dipole transition from the second excited ionic state B to the predissociating state C, whilst recollision excitation was shown to not play a role. The strong field induced coupling of the ionic states B and C could thus be controlled by the laser polarisation.

The results obtained in this thesis further the understanding of population dynamics of cation states in strong field processes. This is of special interest for extending molecular strong field physics to the study of electronic degrees of freedom and their coupling to the nuclear motion.

Acknowledgments

This Ph.D. project would have never been possible without the massive support, advice and motivational speeches that I received on the way.

First of all, I would like to thank my supervisor Jon Marangos. Not only did he dig through a mountain of paperwork to get me this Ph.D. project, he is also one of the main reasons why I had so much (bloody!) fun doing it. It was simply impossible to resist his never ending enthusiasm for physics and his great sense of humour, even in the face of lasers blowing up minutes before a crucial experiment. I felt that he gave me a lot of freedom to develop my own ideas and plan large parts of my project independently and I am extremely grateful for the confidence he put in me. This was exactly what I had hoped for in a research degree and I cannot thank Jon enough for making it such an enriching experience, both professionally and personally.

I would also like to thank the other professors of the laser consortium: Leszek Frasinski, Roland Smith and John Tisch. Be it a replacement vacuum pump, detector or help with theoretical concepts, I could always count on their help. I especially have to thank Leszek for teaching me the ways of the time-of-flight spectrometer and for having the finest moustache around. Special thanks also go to the visiting professor Rashid Ganeev for being a great drinking accomplice and advisor.

Many thanks also go to Misha Ivanov and Olga Smirnova at the Max Born Institute in Berlin. Without their ideas and advice, it would have been impossible to cut through the jungle of excited states and recolliding electrons in CO_2.

However, the fact that my experiments got done and even lead to some meaningful results is mostly due to a group of heroic postdocs from the Red Dragon Lab. They taught me everything I know. Omair Ghafur showed me how cool molecules are (oh yes, molecular-beam-pun intended) and that you need to put on the 'Mission Impossible' theme when you handle the most expensive equipment. I cannot thank Thomas Siegel enough for all his hohehoheho (French for 'awesomeness'). He taught me how to use and (most importantly) fix the laser and his pragmatic approach and incredible talent at planning and building experimental setups were a massive help and inspiration. Amelle Zaïr was the best lab boss/mum I could have asked for. Whenever I got stuck in my project she was there to support me and her trust was invaluable for the progress of my work. Last but not least, I would like to thank Sébastien Weber for being the most awesome lab partner/daddy. Working with him was so much fun and his ideas, motivation and magic molecular alignment matlab code have pushed my work immensely. Not to mention the liquid nitrogen cooled beer that got us through those long experimental runs in Denmark.

Many thanks also to the Attoboys (Felix Frank, Chris Arrell, Tobias Witting and Jarlath McKenna) and Hugo Doyle for all those pieces of advice and borrowed equipment. Special thanks go to Felix for assuming the role of the wise (and cool!) older brother on countless occasions.

I would also like to thank Judith Baylis for being an administrative angel (innit?) and Peter Ruthven, Andy Gregory and Brian Willey from the QOLS workshop for doing all the hard equipment building in the background. Without these incredible people I wouldn't have survived in the world of research and I cannot thank them enough for their work. Special thanks go to Peter for all those genius, quick fixes of whatever was breaking at exactly the most inconvenient time (see Fig. 4.12), but most of all for being the good soul of the basement that makes it feel like a piece of home.

Of course my time in the laser consortium would not have been that much fun without all the other Ph.D. students around. Thank you all for the good times inside and outside the lab, the conversations about life with and without physics and the countless nights we pressed the pub teleportation button. Special thanks go to Simon for all those music-related nights that saved me from lab-induced madness, to Marco for pretending to be a pink flamingo, to Chris for being Henry (sometimes) and to Steffen for being Steffen (always).

Very special thanks go to Elke for being the most amazing and supportive mum (sorry Amelle, she's the real boss). I would not have made it here without her.

Last but not least, I would also like to thank the laser. We had a great time together.

Contents

1 Introduction 1
References 6

2 Molecules in Strong Laser Fields 9
2.1 Atoms in Strong Laser Fields 9
2.1.1 The Strong Field Regime 10
2.1.2 Tunneling Ionisation 13
2.1.3 Free Electron Dynamics and Recollison 16
2.1.4 Nonsequential Double Ionisation 20
2.1.5 High Harmonic Generation 24
2.2 Molecules 25
2.2.1 Electronic Structure 25
2.2.2 Radiative Transitions 30
2.2.3 Field-Dressed States 32
2.2.4 Dissociation Mechanisms 33
2.3 Molecular Tunneling Ionisation 36
2.3.1 Orbital Structure Effects 36
2.3.2 Tunneling Excitation 38
2.4 NSDI in Small Molecules 38
2.4.1 Molecular Structure Effects 40
2.4.2 Laser Induced Alignment Effects 41
2.4.3 NSDI in Aligned Diatomics 41
2.5 Recollision Induced Fragmentation 42
2.5.1 Small Molecules 42
2.5.2 Large Molecules 43
References 45

3 Lasers 51
3.1 Laser Fundamentals 52
3.1.1 Mathematical Description 52
3.1.2 Dispersion Effects 54
3.1.3 Nonlinear Optics 56
3.1.4 Optical Parametric Amplification 57
3.1.5 Optical Kerr Effect 58

3.2 Laser Pulse Production 59
3.2.1 Mode-Locking 60
3.2.2 Chirped Pulse Amplification 62
3.3 Pulse Length Diagnostics 63
3.3.1 Autocorrelation and FROG 64
3.3.2 General SPIDER Technique 64
3.3.3 SPIDER Apparatus 66
3.3.4 NIR SEA-F-SPIDER Apparatus 67
3.4 The Imperial College CPA System 67
3.4.1 Oscillator 67
3.4.2 Amplification 68
3.5 The Imperial College NIR OPA System 70
3.5.1 Seed Generation 70
3.5.2 Amplification 71
References 72

4 Experimental Methods and Setup 75
4.1 Ion Detection Techniques 76
4.1.1 Ion Detection 76
4.1.2 Time-of-Flight Mass Spectrometry: 1D Imaging 78
4.1.3 Velocity Map Imaging: 2D Imaging 82
4.1.4 Further Techniques: 3D Imaging and Beyond 83
4.2 The Imperial College TOFMS 85
4.2.1 Spectrometer Description 85
4.2.2 Acquisition of TOF Spectra 87
4.3 Experimental Setup 87
4.3.1 Vacuum System 87
4.3.2 Pump-Probe Beam Line 92
4.3.3 Data Acquisition 97
4.3.4 Gas Mixing Setup 99
4.4 Molecular Beam 100
4.4.1 Supersonic Gas Jets 101
4.4.2 Cooling Mechanism 102
4.4.3 Skimmed Molecular Beams 104
4.4.4 Seeded Gas Jets 105
4.4.5 Molecular Beam Characterisation 105
4.5 Manipulation of Laser Pulse Parameters 107
4.5.1 Focussing 107
4.5.2 Intensity Estimation 110
4.5.3 Polarisation State 113
4.5.4 Calibration of Polarisation Optics 114
References 117

5 Characterisation and Optimisation of Impulsive Molecular Alignment in Mixed Gas Samples . . . 121
5.1 Background . . . 122
5.1.1 Adiabatic Alignment . . . 122
5.1.2 Impulsive Alignment . . . 123
5.1.3 Adiabatic Versus Impulsive Alignment . . . 124
5.1.4 Detecting and Characterising Molecular Alignment . . . 125
5.2 Theory of Impulsive Alignment . . . 126
5.2.1 Theoretical Model . . . 127
5.2.2 Thermal Ensemble . . . 130
5.2.3 Numerical Simulations . . . 131
5.3 Impact of the Rotational Temperature . . . 131
5.4 Experiment . . . 133
5.4.1 Setup and Procedure . . . 133
5.4.2 Revival Identification . . . 135
5.5 Results . . . 138
5.5.1 N_2 . . . 138
5.5.2 CO_2 and O_2 . . . 139
5.5.3 Discussion . . . 139
5.6 Matching Theory to Experiment . . . 140
5.7 Conclusion . . . 143
References . . . 144

6 Multichannel Contributions in Nonsequential Double Ionisation of CO_2 . . . 147
6.1 Background . . . 147
6.1.1 Strong Field Ionisation . . . 148
6.1.2 Nonsequential Double Ionisation . . . 148
6.2 Analytical Model for Estimating the Visibility of Quantum Interference . . . 151
6.2.1 Single Molecule . . . 152
6.2.2 Convolution With Molecular Alignment Distribution . . . 154
6.3 Experiment . . . 155
6.3.1 Setup and Procedure . . . 156
6.3.2 Data Analysis . . . 158
6.4 NSDI in Ar . . . 159
6.4.1 Results . . . 159
6.4.2 Discussion . . . 161
6.5 Intensity Dependence of NSDI in CO_2 . . . 164
6.5.1 Results . . . 164
6.5.2 Discussion . . . 165
6.6 Ellipticity Dependence of NSDI in CO_2 . . . 165
6.6.1 Results . . . 165
6.6.2 Discussion . . . 167

6.7 Polarisation Dependence of NSDI in CO_2 168
6.7.1 Results. 168
6.7.2 Discussion 169
6.7.3 Multichannel NSDI 171
6.8 Conclusion 173
References 173

7 Strong Field Control of Dissociative Excitation in CO_2^+ 175
7.1 Background 175
7.1.1 Predissociation Mechanism 176
7.1.2 Branching Ratio 178
7.1.3 Excitation Pathways 178
7.2 Experiment 180
7.2.1 Setup and Procedure 180
7.2.2 Collection of Fragment Ions 181
7.3 Dissociative Excitation to $CO^+_{(1,0)}$ 182
7.3.1 Intensity Dependence. 183
7.3.2 Ellipticity Dependence. 184
7.3.3 Polarisation Dependence 186
7.4 Dissociative Excitation to $O^+_{(1,0)}$ 189
7.4.1 Intensity Dependence. 189
7.4.2 Ellipticity Dependence. 190
7.4.3 Polarisation Dependence 192
7.5 Branching Ratio 195
7.5.1 Intensity Dependence. 196
7.5.2 Polarisation Dependence 197
7.6 Conclusion 198
References 199

8 Conclusion 201
8.1 Chapter 5: Characterisation and Optimisation of Molecular Alignment 201
8.2 Chapter 6: Recollision Induced Double Ionisation 202
8.3 Chapter 7: Strong Field Induced Dissociation 204
Reference 205

Chapter 1
Introduction

Molecular physicists face a dilemma. They'd love to study the microscopic world inside molecules and shed light onto the fundamental dynamics that drive chemical and biological processes. Unfortunately they live in a macroscopic world. Because of this, molecular physicists need suitable experimental tools to access and manipulate the molecular properties they are interested in.

In this introduction I will provide a first glimpse at those molecular properties that are of interest for this thesis and the chosen experimental tools to access them. This should provide the reader with a roadmap that marks the most important points of departure for the conducted research.

Molecular physicists typically approach their work with a so-called bottom-up approach. This means that they attempt to understand the behaviour of molecules by investigating the properties and interactions of the components that comprise them. These components are the positively charged nuclei and negatively charged electrons that make up the individual atoms that in turn are bound together to form the molecule. Here, Sect. 2.2.1 provides a brief introduction to molecular physics with all basic terms and mechanisms relevant to this thesis.

Reasearch in molecular physics can then be divided into two strands: structural and dynamical. The first studies the specific properties of a given molecule, for instance the geometric arrangement of its atomic nuclei and electron distribution (which is directly linked to the molecule's so-called *orbital*) and the resulting electronic, vibrational and rotational energy levels (which are associated with the molecule's so-called *quantum states*). The second strand investigates how these properties change in time; as part of a chemical reaction for example. The relevant time scales of such molecular dynamics are essentially determined by the motion of the involved molecular components. Due to the comparatively large mass of the atomic nuclei the associated dynamics such as vibrations, rotations or changes of the molecular structure typically cover durations from hundreds of picoseconds ($1\,\text{ps} = 10^{-12}\,\text{s}$) down to a few femtoseconds ($1\,\text{fs} = 10^{-15}\,\text{s}$). Here, the fastest vibrational period can be found in an H_2 with about 8 fs. Due to these extremely short time scales, the associated field of research is often called *ultrafast molecular dynamics*.

M. Oppermann, *Resolving Strong Field Dynamics in Cation States of CO_2 via Optimised Molecular Alignment*, Springer Theses, DOI: 10.1007/978-3-319-05338-7_1,

However, as an electron is about a thousand times lighter than the lightest atomic nucleus (a proton), the associated dynamics are accordingly faster. The molecule's internal electron motions thus take place on the attosecond time scale (1 as $= 10^{-18}$ s) and is related to phenomena such as electron charge migration in molecules and orbital changes during a molecular energy level transition, for example (for a review of the emerging field of attosecond physics see [1]).

Here, one should also note that there is an intermediate regime in molecular dynamics, where the motion of the nuclei and electrons cannot be studied separately. This is the case, when these molecular components interact with each other and is often referred to as a coupling of the molecule's nuclear and electronic degrees of freedom. Here, energy can be transferred between the interacting particles, such that a molecular vibration may enable an internal electronic exciation. However, the interaction of the particles increases the complexity of such phenomena and makes their investigation and description very challenging.

It is at this point, where the current frontier of ultrafast molecular dynamics research becomes apparent: the description and control of ever faster molecular phenomena in increasingly complex molecules. Throughout Chap. 2, it will become clear that this not only requires the development of novel theoretical and numerical tools, but also new experimental techniques and high quality data for benchmarking the theory. This thesis is based on this latter experimental perspective.

In this context, lasers have become a popular and equally successful experimental tool for molecular physicists. This is because lasers are an astonishingly precise macroscopic tool for inducing and manipulating microscopic quantum state distributions, due to their coherent, collimated, intense and potentially frequency-tunable flux of photons. This is outlined in more detail in Sects. 2.1 and 2.2, which discuss the basic mechanisms underlying light-matter interactions. Their study using lasers as light sources is called *laser spectroscopy*. Chapter 3 then provides a basic introduction to laser physics and presents the basic architecture and working principle of the laser systems employed for this thesis. Whilst the basic properties of laser light are thus at the heart of molecular laser spectroscopy, three further developments in this field form the context of this thesis.

The first development is the application of pulsed lasers for the time resolution of laser spectroscopy techniques. The use of laser pulses enables a time-sequencing of the induced dynamics within so-called pump-probe setups. Here, a first laser pulse (the pump) triggers a molecular process, like a photoexcitation to a so-called repulsive molecular state that leads to the breaking of a molecular bond. A second laser pulse (the probe) then probes the molecular properties as a function of time delay after the pump pulse, for instance the distance of the molecular fragments as a function of time after the pump. In this example, the nuclear motion during a chemical reaction is mapped out as a function of time. However, this requires that the probe pulse is shorter than the typical time scale of the motion of the atoms in a molecule. The application of femtosecond laser pulses with durations around 20 fs thus created the field of femtochemistry, aiming at the time-resolution of the nuclear motion in molecules [2].

The second development is the use of high power femtosecond laser pulses for entering a new regime of light-molecule interaction, the so-called strong field regime (see Sect. 2.2.1). It is marked by the onset of optical field ionisation, which takes place when the laser electric field becomes comparable in strength to the binding Coulomb field in the molecule. The Coulomb potential of the molecule is thus suppressed and a finite potential barrier formed that allows an electron to escape from the molecule via tunneling (see Sect. 2.2.1). This ionisation event is the fundamental process in strong field physics. Whilst being a research topic in its own right, tunneling ionisation also represents an ultrafast (less than a few tens of attoseconds [3, 4]), highly nonlinear response of the molecule to an electric field and can thus be used as a probing process for molecular properties. At the same time, the ionisation step triggers molecular dynamics that can be studied and controlled through the driving laser field. This approach has become an excellent tool for imaging molecular structure and internal dynamics. It is built on the high degree of control of the driving laser field parameters achieved over the last two decades [1, 5]. Pulse durations in the near-infrared (NIR) from 30 fs down to below 4 fs [6] now routinely provide electric field gradients steep enough to achieve the peak intensities on the order of $10^{14}\,\mathrm{Wcm^{-2}}$ that are required for observing strong field phenomena.

This development has informed the application of ultrafast strong field processes for probing molecular dynamics, with the aim to extend this study to the investigation of the electronic degrees of freedom and their coupling to the nuclear degrees of freedom. Recent examples for achieving this include the use of high harmonic generation [7], time-resolved photoelectron spectroscopy [8] and strong field ionisation [9], to name but a few highlights from the last two years. Rather recently, the control of the carrier envelope phase and the production of pulse durations in the attosecond regime (for a recent review of the required experimental tools see for example [10]) have even opened routes towards imaging the internal electron dynamics in molecules in a weak field regime [1]. Here, Sects. 2.3, 2.4 and 2.5 provide a review of those parts of molecular strong field physics relevant to this thesis.

The third development is the introduction of laser induced impulsive molecular alignment to resolve the molecular structure in strong field processes (for a review see [11]). Molecules in gas phase are randomly aligned within the laboratory frame of reference. This implies that the interaction of the molecular ensemble with a linearly polarised laser field is averaged over all possible orientations of the molecule with respect to the laser polarisation. This makes it impossible to resolve the effect of the molecular structure on the studied laser induced processes.

However, with the use of laser induced impulsive alignment, one may impose a preferred direction of alignment on the molecular ensemble within the laboratory frame. In this way its interaction with the laser takes place in the molecular frame of reference and the role of the molecular structure may be resolved. As an example, one may now record the ionisation rate in a given linear molecule as a function of the angle between the linear probe laser polarisation and the molecular axis. Here, the angular distribution of photoelectrons would be a direct mapping of the ionised orbital of the probed molecule.

Impulsive molecular alignment is presented in full detail in Chap. 5. Briefly, a relatively weak (10^{13} Wcm^{-2}) femtosecond laser pulse kicks a linear molecule and forces it to rotate. Typical small diatomic molecules like N_2 have rotational periods of several picoseconds, so the interaction of the laser pulse is much shorter than the time-scale of the induced rotation. The laser pulse thus induces a coherent superposition of rotational states of the molecule creating a rotational wavepacket that is initially aligned along the laser pulse polarisation. Due to the imposed phase relationships between the coupled rotational states, the wavepacket subsequently de- and rephases periodically at times that are characteristic to the molecule. Constructive interference of the rotational states is associated with alignment of the wavepacket along the initial laser pulse polarisation and thus allows the probing of a molecular ensemble fixed in the laboratory frame of reference under field free conditions.

Through the above developments, the control of strong field induced dynamics in small molecules on the femtosecond time scale continues to be a fruitful branch of research; also for establishing the experimental and theoretical tools required to go towards ever faster molecular dynamics and more complex molecular systems. This is the context of the experiments presented in this thesis. They aim at resolving the role of the molecular structure in strong field induced processes in CO_2 by the use of optimised impulsive molecular alignment.

CO_2 was chosen as a target molecule for several reasons. It is a triatomic molecule and thus the smallest possible polyatomic system. This implies that all fundamental molecular processes, such as vibration, bending and internal couplings can play a role in the probed dynamics, whilst the complexity of the molecule is kept as small as possible. Therefore CO_2 is a suitable model system to study ultrafast dynamics in the presence of a strong laser field.

Additionally, the choice of CO_2 also offers advantages from a practical point of view. Firstly a rather large body of work is available on the tunneling ionisation of CO_2 [12–14]. Secondly, the dissociation of the molecular ion CO_2^+ is relatively well understood [15]. Thirdly, the resulting fragmentation channels could be identified and isolated rather easily in the used experimental setup due to their lack of kinetic energy release [16]. Two processes were then investigated: the second ionisation and the dissociation mechanism of the molecular ion CO_2^+. Both processes are driven by the initial tunneling ionisation of the neutral molecule. Through the use of molecular alignment, the orbital symmetries of the involved molecular ionic states could be resolved, such that the internal dynamics in the form of ionisation and excitation pathways were tracked down.

The ionisation of CO_2^+—or double ionisation of CO_2—can proceed in two ways. Firstly, the ion can be field-ionised. Here, the first and second ionisation event are independent of each other. In the second mechanism, CO_2 is ionised and the ejected electron driven into the continuum by the strong laser field. Due to its oscillatory motion, the free electron can be driven back to the parent ion and may excite or impact ionise it when they collide. This process is called laser driven electron recollision [17] and is attracting wide attention from strong field physicists as it represents an ultrafast process that can be fully controlled by the driving laser field parameters (see Sect. 2.1.3). In fact, the excursion time of the recolliding electron is about 3/4 of the laser field's oscillation period and can thus be tuned by its wavelength. The impact energy is given by the electric field strength and total acceleration time and hence can be tuned by the intensity and wavelength, respectively. Furthermore, the electron trajectory is confined to the plane of polarisation of the laser field. Using molecular alignment, the electron's recollision angle with respect to the molecular axis can thus be controlled by the laser polarisation.

This control perspective is especially interesting as electron recollision allows for inducing and probing ultrafast molecular dynamics combining sub-Ångstrom spatial with sub-femtosecond time resolution [18]. Here, the spatial resolution is related to the wavelength of the electron that is on the order of hundreds of picometers ($1\,\mathrm{pm} = 10^{-12}\,\mathrm{m}$), whereas the time resolution is rooted in the excursion time of the electron in the continuum. This allows to time-resolve the sub-cycle dynamics in the molecule that take place between the initial tunneling ionisation and recollision event. These aspects have lead to milestone experiments tracking down the nuclear motion in molecules [19, 20], structural changes during chemical reactions [21] or the tomographical imaging of molecular orbitals [22], to name but a few examples. In the experiment presented in this thesis, recollision induced double ionisation was controlled via increasing the driving laser wavelength and impulsive molecular alignment to angularly resolve the mechanism. This allowed for the identification of the intermediate ionic states involved in the double ionisation mechanism of CO_2 and thus suggests a novel way for employing inelastic electron recollisions as an ultrafast spectroscopy technique.

The dissociation of CO_2^+ requires its excitation to a repulsive state. In the strong field regime, this is possible via several pathways. Firstly, the tunneling ionisation of CO_2 may take place from a low lying orbital, which would leave the parent ion in an excited electronic state. In addition, inelastic recollisions [16] and laser induced photo-excitation may facilitate the population of excited ionic states. The experiments presented in this thesis firstly aim at identifying the contributions from the above channels to the dissociation mechanism by recording the resulting fragmentation yield as a function of different laser parameters such as intensity, ellipticity and polarisation. As the above excitation pathways proceed via different ionic states, their angular dependence in the molecular frame is related to the associated molecular orbitals. It was found that the fragmentation mechanism of CO_2^+ is dominated by a strong field induced coupling between its second and third excited state. This coupling and thus the fragmentation yield could be controlled by the laser polarisation with respect to the molecular axis and a signal suppression of up to 70 % was achieved. This result is especially interesting from a theoretical point of view as it offers the possibility to study the ultrafast population transfer between electronically coupled states in the strong field regime. Theoretical tools that successfully describe such complex scenarios are needed for extending strong field spectroscopy to even more challenging processes such as charge migration or the coupling of electronic and nuclear degrees of freedom in biological systems.

In order to contextualise the research conducted for this thesis, I have structured its contents as follows. In Chap. 2, the strong field regime and its associated phenomena are introduced and applied to the case of small linear molecules. This includes a brief introduction to molecular physics and the discussion of previous experimental research on double ionisation and strong field induced dissociation relevant to this thesis. In Chaps. 3 and 4, the laser system and the experimental setup employed for the experiments are presented and discussed in detail. Chapter 5 then introduces the technique of laser induced impulsive molecular alignment from a theoretical and experimental perspective. In particular, a novel procedure for characterising the quality of the resulting molecular alignment distribution in the used experimental setup was developed. This was applied to optimise molecular alignment in mixed gas samples that could then be used in the strong field experiments. Chapter 6 then presents the experiment on recollision induced double ionisation in CO_2, whilst Chap. 7 focusses on strong field induced dissociation of CO_2^+. In the last chapter, the results presented in this thesis are summarised. This leads to the discussion of their contribution to the scientific community, possible improvements and the motivation of new experiments.

References

1. F. Krausz, M. Ivanov, Attosecond physics. Rev. Mod. Phys. **81**, 163 (2009)
2. A.H. Zewail, Laser femtochemistry. Science **242**, 1645–1653 (1988)
3. P. Eckle, A.N. Pfeiffer, C. Cirelli, A. Staudte, R. Dörner, H.G. Muller, M. Büttiker, U. Keller, Attosecond ionization and tunneling delay time measurements in helium. Science **322**, 1525–1529 (2008)
4. D. Shafir, H. Soifer, B.D. Bruner, M. Dagan, Y. Mairesse, S. Patchkovskii, M.Y. Ivanov, O. Smirnova, N. Dudovich, Resolving the time when an electron exits a tunnelling barrier. Nature **485**, 343–346 (2012)
5. T. Brabec, F. Krausz, Intense few-cycle laser fields:frontiers of nonlinear optics. Rev. Mod. Phys. **72**, 545 (2000)
6. T. Witting, F. Frank, C.A. Arrell, W.A. Okell, J.P. Marangos, J.W.G. Tisch, Characterization of high-intensity sub-4-fs laser pulses using spatially encoded spectral shearing interferometry. Opt. Lett. **36**(9), 1680–1682 (2011)
7. H.J. Wörner, J.B. Bertrand, B. Fabre, J. Higuet, H. Ruf, A. Dubrouil, S. Patchkovskii, M. Spanner, Y. Mairesse, V. Blanchet, E. Mvel, E. Constant, P.B. Corkum, D.M. Villeneuve, Conical intersection dynamics in NO_2 probed by homodyne high-harmonic spectroscopy. Science **334**, 208–212 (2011)
8. P. Hockett, C.Z. Bisgaard, O.J. Clarkin, A. Stolow, Time-resolved imaging of purely valence-electron dynamics during a chemical reaction. Nat. Phys. **7**, 612–615 (2011)
9. A.E. Boguslavskiy, J. Mikosch, A. Gijsbertsen, M. Spanner, S. Patchkovskii, N. Gador, M.J.J. Vrakking, A. Stolow, The multielectron ionization dynamics underlying attosecond strong-field spectroscopies. Science **335**, 1336–1340 (2012)
10. F. Frank, C. Arrell, T. Witting, W. A. Okell, J. McKenna, J.S. Robinson, C.A. Haworth, D. Austin, H. Teng, I.A. Walmsley, J.P. Marangos, J. W.G. Tisch, Invited review article: technology for attosecond science, Rev. Sci. Instrum. **83**(7), 071101–071101–23 (2012)
11. H. Stapelfeldt, T. Seideman, Colloquium: aligning molecules with strong laser pulses. Rev. Mod. Phys. **75**, 543 (2003)
12. D. Pavicic, K.F. Lee, D.M. Rayner, P.B. Corkum, D.M. Villeneuve, Direct measurement of the angular dependence of ionization for N_2, O_2, and CO_2 in intense laser fields. Phys. Rev. Lett. **98**(24), 243001 (2007)

13. O. Smirnova, Y. Mairesse, S. Patchkovskii, N. Dudovich, D. Villeneuve, P. Corkum, M.Y. Ivanov, High harmonic interferometry of multi-electron dynamics in molecules. Nature **460**(7258), 972–977 (2009)
14. L. Torlina, M. Ivanov, Z.B. Walters, O. Smirnova, Time-dependent analytical r-matrix approach for strong-field dynamics II many-electron systems. Phys. Rev. A **86**, 043409 (2012)
15. M.T. Praet, J.C. Lorquet, G. Raseev, Unimolecular reaction paths of electronically excited species iv. the $^2\sigma_g^+$ state of CO_2^+. J. Chem. Phys. **77**(9), 4611–4618 (1982)
16. J. McKenna, M. Suresh, B. Srigengan, I.D. Williams, W.A. Bryan, E.M.L. English, S.L. Stebbings, W.R. Newell, I.C.E. Turcu, J.M. Smith, E.J. Divall, C.J. Hooker, A.J. Langley, J.L. Collier, Rescattering-enhanced dissociation of a molecular ion. Phys. Rev. A **74**, 043409 (2006)
17. P.B. Corkum, Plasma perspective on strong field multiphoton ionization. Phys. Rev. Lett. **71**, 1994–1997 (1993)
18. M. Lein, Molecular imaging using recolliding electrons. J. Phys. B: At. Mol. Opt. Phys. **40**(16), R135–R173 (2007)
19. S. Baker, J.S. Robinson, C.A. Haworth, H. Teng, R.A. Smith, C.C. Chiril, M. Lein, J.W.G. Tisch, J.P. Marangos, Probing proton dynamics in molecules on an attosecond time scale. Science **312**(5772), 424–427 (2006)
20. H. Niikura, F. Legare, R. Hasbani, M.Y. Ivanov, D.M. Villeneuve, P.B. Corkum, Probing molecular dynamics with attosecond resolution using correlated wave packet pairs. Nature **421**, 826–829 (2003)
21. H.J. Worner, J.B. Bertrand, D.V. Kartashov, P.B. Corkum, D.M. Villeneuve, Following a chemical reaction using high-harmonic interferometry. Nature **466**(7306), 604–607 (2010)
22. J. Itatani, J. Levesque, D. Zeidler, H. Niikura, H. Pepin, J.C. Kieffer, P.B. Corkum, D.M. Villeneuve, Tomographic imaging of molecular orbitals. Nature **432**(7019), 867–871 (2004)

Chapter 2
Molecules in Strong Laser Fields

The purpose of this chapter is to provide the background and context for interpreting the experiments presented in this thesis. In particular, they aim at understanding the internal dynamics of CO_2 induced by a strong laser field. For a scientist (and especially for a Ph.D. student), this is a rather worrying prospect. Due to the complexity of molecular systems there are usually several mechanisms contributing to one particular outcome. This complicates the interpretation of the experimental results. Often, quantitative numerical simulations can help resolve this issue. However the strong laser field distorts the molecular system significantly and only adds to its complexity. This renders quantitative calculations a challenging problem, usually requiring a (second?) Ph.D. in computational physics. This thesis attempts to escape from this dilemma by measuring the molecular dynamics as a function of many different laser parameters. In this way, the dominating mechanism may be identified, enabling a qualitative understanding of the results. This chapter is thus focussed on introducing the studied strong field processes (ionisation, double ionisation and dissociation) and their dependence on the laser parameters (intensity, wavelength, polarisation and ellipticity). Emphasis is placed on available experimental results where possible.

2.1 Atoms in Strong Laser Fields

As molecules are composed of atoms, it is helpful to start with their response to a strong electric field. When near-infrared radiation is used as in the experiments discussed in this thesis, the corresponding *strong field regime* is roughly characterised by the onset of *ionisation*—the removal of an electron from the atom. In the following sections, the possible ionisation pathways are discussed in more detail, before looking at the interaction of the freed electron with the remaining laser field.

M. Oppermann, *Resolving Strong Field Dynamics in Cation States of CO_2 via Optimised Molecular Alignment*, Springer Theses, DOI: 10.1007/978-3-319-05338-7_2,

2.1.1 The Strong Field Regime

The ionisation potential I_p of an atom is defined as the energy required to lift its most loosely bound electron into the continuum (a bound-continuum transition). When irradiated by light of sufficiently high frequency ω, this is possible via a single-photon transition [1]. Yet, ionisation is also possible in lower frequency fields. Here, ionisation takes place via the absorption of n photons of energy $\hbar\omega$, such that $(n-1)\hbar\omega < I_p < n\hbar\omega$. This was first observed in 1965 by Voronov and Delone, involving a seven-photon transition in Xenon [2]. The corresponding multiphoton ionisation (MPI) rate can be obtained by treating the field as a small perturbation with respect to the Coulomb potential of the atom [3]. The associated theoretical framework is called lowest order perturbation theory (LOPT) and results in an intensity scaling that takes the form of a power law:

$$\Gamma_n^{(MPI)} = \sigma_n I^n \tag{2.1}$$

where σ_n is the n-photon absorption cross-section and I the light intensity. However, with increasing intensity, higher order terms become significant requiring an extension of LOPT. In 1979 Agostini et al. then observed MPI transitions with additional excess or *above threshold* photons [4]. This is illustrated in part (a) of Fig. 2.1. These continuum-continuum transitions increase the freed electron's kinetic energy. The MPI rate of an $(n+s)$-photon transition thus scales in intensity to the power of (n+s). This has been verified in an intensity range up to $10^{12}\,\mathrm{W\,cm^{-2}}$ [5]. However, the observation of high order above threshold ionisation usually requires intensities on the order of $10^{13}\,\mathrm{W\,cm^{-2}}$, due to the fast decrease of the absorption cross-section with the photon number. Here, the oscillatory laser field becomes strong enough to couple to the atomic states [6]. This effect is called the AC-Stark shift and becomes stronger with increasing intensity and the more loosely the electron is bound in the atom. One may estimate an upper limit of the associated energy level shift by treating the most weakly bound electron as freely oscillating in the laser field. This approximation is appropriate for atomic Rydberg states, for example. It thus approximately acquires the cycle averaged quiver energy of a free electron in an AC-field, called the ponderomotive potential U_p. Given a laser field of the form $E_L(t) = E_L \sin(\omega_L t)$, the electron's acceleration then is $a(t) = \dot{v}(t) = \frac{e}{m_e} E_L(t)$, such that [3]:

$$U_p[eV] = \left\langle \frac{1}{2} m_e v(t)^2 \right\rangle = \frac{(eE_L)^2}{4 m_e \omega_L^2} \approx 9.33 \times 10^{-14} I[\mathrm{W\,cm^{-2}}]\lambda^2[\mu\mathrm{m}] \tag{2.2}$$

where m_e is the electron's mass and λ the laser wavelength. The effective ionisation potential is thus increased by this energy (see part (a) of Fig. 2.1), leading to a reduction of the ejected electron's kinetic energy by the same amount:

$$E_{kin} \approx (n+s)\hbar\omega - (I_p + U_p) \tag{2.3}$$

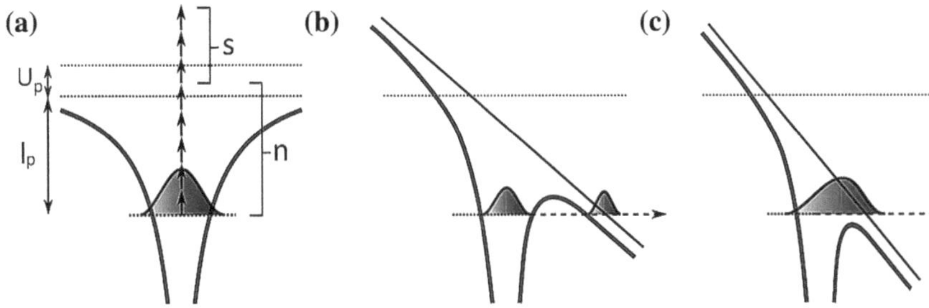

Fig. 2.1 Ionisation scenarios in the strong field regime: multiphoton ionisation (**a**), tunneling ionisation (**b**) and over-the-barrier ionisation (**c**)

where s is the number of excess photons. Once the ponderomotive energy becomes larger than the photon energy, entire peaks in the low kinetic energy region of the above-threshold ionisation (ATI) spectra are suppressed [7]. For this, the laser field cannot be treated as a perturbation anymore and LOPT treatment together with the multiphoton picture begins to fail. Approaching laser intensities on the order of $10^{14}\,\mathrm{W\,cm^{-2}}$ this also becomes apparent through another feature in atomic ATI. Here, Paulus et al. were the first to observe a plateau in the ATI spectra, associated with an almost constant ionisation probability over an extended range of high order ATI transitions [8, 9]. As LOPT leads to a strong exponential decrease of the ionisation cross-section with the ATI order, perturbation theory fails and a non-perturbative treatment is required. This was first realised by Keldysh in 1965 via a semiclassical model that treats the laser field classically [10]. Due to its non-perturbative interaction with the atom's Coulomb potential a barrier of finite width is formed. This allows the build-up of a tunneling current, resulting in field ionisation. This mechanism is thus called *tunneling ionisation* (TI) and is illustrated in part (b) of Fig.2.1. MPI and TI represent the limiting cases of ionisation in a strong laser field, however there is no sharp transition between them at a particular intensity.

Keldysh developed an adiabaticity parameter γ that allows for a rough classification of the ionisation scenario as a function of the ionisation potential and laser field parameters. For this, one may assume that the electron is bound in an infinitely thin potential well of depth I_p, which corresponds to its ionisation potential. It is distorted by the time-averaged electric field of amplitude E_L, forming a triangular barrier. The resulting tunnel length is $l = I_p/(eE_L)$ and the kinetic energy the electron carries after tunneling equals the ionisation potential. Hence the tunneling frequency ω_t can be estimated via the resulting tunneling period T_t and velocity v_t:

$$\omega_t = \frac{2\pi}{T_t} = \frac{2\pi v_t}{l} = \frac{2\pi e E_L}{\sqrt{I_p m_e/2}} \approx \frac{eE_L}{\sqrt{2 I_p m_e}} \tag{2.4}$$

The above expression suggests that the tunneling process only depends on the laser intensity, but once $\omega_t \ll \omega_L$ a tunneling current cannot build up during one laser cycle. In this case, the ionisation becomes laser frequency dependent, hence approaching

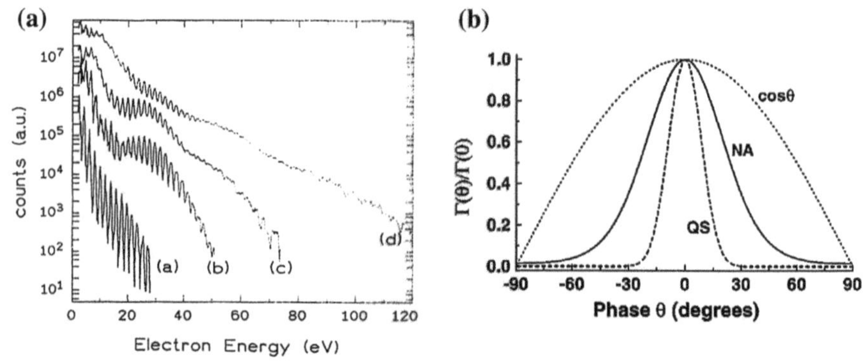

Fig. 2.2 **a** ATI spectra in Ar using 40 fs, 630 nm pulses. The employed intensities are 0.6 (*a*), 1.2 (*b*), 2.4 (*c*) and $4.4 \times 10^{14}\,\mathrm{W\,cm^{-2}}$ (*d*). The *right panel* **b** displays calculated instantaneous ionisation rates $\Gamma^{(TI)}$ in He as a function of the laser field phase. The employed laser parameters are $\lambda = 780\,\mathrm{nm}$ and $I = 5 \times 10^{13}\,\mathrm{W\,cm^{-2}}$. Both quasistatic (*QS*) and nonadiabatic (*NA*) rates are shown, together with the cosine shape of the laser field (Reprinted with permission from [9] and [13]. Copyright (1994) and (2001) by the American Physical Society)

the MPI scenario. The Keldysh parameter is thus defined by comparing tunneling and laser frequency:

$$\gamma = \frac{\omega_L}{\omega_t} = \frac{\omega_L\sqrt{2I_p m_e}}{eE_L} = \sqrt{\frac{I_p}{2U_p}} \tag{2.5}$$

such that for $\gamma \ll 1$ TI and for $\gamma \gg 1$ MPI dominates. $\gamma \approx 1$ is an intermediate regime where high order perturbation theory and non-perturbative methods may be appropriate. The tunneling regime is often referred to as the pure strong field regime and will be discussed in more detail in the Sect. 2.1.2. The transition between the regimes can be illustrated via ATI spectra recorded at different intensities as shown in part (a) of Fig. 2.2. Starting from an exponentially decaying spectrum (pure MPI), an additional plateau is formed at higher intensity (intermediate regime). At even higher intensities the peak structure vanishes and multiphoton effects are absent (pure TI).

Even though not included in the Keldysh parameter, strong field ionisation implicitly requires short laser pulses. This is due to the fact that for a longer pulse, the electric field gradient is not steep enough. This means that the increase of the peak field is too slow, such that the ground state population is depleted before the maximum peak intensity of the pulse is reached. In particular for pulses in the visible or NIR spectral range, durations <10 fs are required to access the pure strong field regime [11, 12]. In this context, the saturation intensity I_{sat} is defined as the effective intensity were the ground state population is fully depleted. In this saturation regime, the ionisation rate remains constant with increasing intensity.

2.1.2 Tunneling Ionisation

There is a wealth of theoretical approaches to TI that have entertained theoretical physicists since Keldysh's early work. One branch has developed from extensions to Keldysh's theory [14, 15] into the so-called Strong Field Approximation (SFA) [16]. Based on a semiclassical treatment, Perelomov, Popov and Terent'ev developed another tunneling model (PPT) [17–19] that was further extended by Ammosov, Delone and Krainov and is now known as ADK-theory [20]. Luckily, all tunneling models lead to the same qualitative exponential ionisation rate that is observed experimentally (for a comparison see for example [21]). In atomic units (where the elementary charge, electron mass, reduced Planck's constant and Coulomb's constant are set to unity), the tunneling rate can then be expressed as follows:

$$\Gamma^{(TI)} \propto \exp\Big(-\frac{2}{3}\frac{(2I_p)^{3/2}}{E_L}\Big) \tag{2.6}$$

For this thesis, SFA is of particular interest as it was developed for also including the free electron's motion in the laser field after TI, which is described in the following Sect. 2.1.3. The SFA comprises a set of approximations, in order to transform the TI event into an analytically solvable problem:

1. *Single Active Electron* (*SAE*) Only one electron is considered in the ionisation event, while the interaction with the remaining electronic states is neglected [12, 22]. Furthermore, only the ground state is taken into account in the evolution of the system. This is reasonable as long as multi-electron excitations are significantly less probable than single-electron excitations, which is usually the case for noble gas atoms and small molecules in NIR fields [12].
2. *Dipole Approximation* (*DA*) The electric field is spatially homogeneous on the scale of the atom, which is treated as an electric dipole. This implies that the laser wavelength is significantly longer than the atomic dimensions, which is justified for NIR radiation and small molecules.
3. *Strong Field Approximation* (*SFA*) The laser field is included exactly and treated classically, while the Coulomb potential acts as a perturbation. The zeroth-order SFA thus completely neglects the potential. Due to the exponential decay of the TI rate with decreasing field strength (see Eq. 2.6), ionisation only takes place efficiently near the peak of the field (see part (a) of Fig. 2.2). In the strong field regime, the laser field is thus much stronger than the Coulomb potential at the moment of ionisation. This is the main justification for the SFA.
4. *Quasistatic Approximation* (*QSA*) The laser field oscillations are slow compared to the atomic response. For this, the electric field is treated as a slowly varying DC field. The frequency dependence of the ionisation rate thus vanishes. Strictly speaking this is only justified in the pure strong field regime, where $\gamma \ll 1$ [13]. However, in the guise of PPT and ADK models, the QSA has been verified experimentally for rare gas atoms [21] and a range of small molecules [23] up to $\gamma \approx 3$. Nevertheless, this is not sufficient if one is interested in sub-lasercycle dynamics

in this intermediate regime. This is the case for measurements involving the CEP of few-cycle pulses, for example. For this, Yudin and Ivanov have developed an extension of the PPT model for calculating sub-cycle, time-dependent ionisation rates independent of the value of γ [13] (see part (b) of Fig. 2.2). The model does not rely on the QSA and is referred to as non-adiabatic tunneling.

Many strong field effects can be understood within the SFA framework. Key findings in the context of this thesis are summarised below.

Properties of the tunneled electron

In the simplest approximation, the electron is assumed to appear in the continuum at position zero with zero momentum. This omits the finite width of the tunneling barrier and initial momentum spread of the tunneled electron. However, as initial conditions for a classical propagation of the free electron in the laser field they reproduce surprisingly many experimental observations (see Sect. 2.1.3). In a more rigorous quantum mechanical treatment there is not a single final continuum state—in this case associated with zero momentum. Rather, during ionisation some population from the electronic ground state is transferred to a set of continuum states, each associated with a particular electron momentum. The free electron is thus described as the superposition of all possible momentum states in the continuum, weighted by their respective transition probability. The resulting wavepacket's transverse velocity distribution is relevant for this thesis and can be expressed in atomic units as follows [24]:

$$\psi(v_\perp) = \psi(0) \exp\left(-\frac{v_\perp^2}{2} \frac{\sqrt{2I_p}}{E_L} \right) \tag{2.7}$$

where $\psi(0)$ is the normalisation factor and $|\psi(v_\perp)|^2$ is the probability of the electron to have the transverse velocity $v_\perp$ after tunneling. The distribution is centered at zero transverse momentum, justifying the classical initial conditions. In longitudinal direction, the emerging wavepacket has a more complicated shape. Also, its centre is shifted towards the laser field's acceleration direction [24], with $v_\parallel^2/2 \ll I_p$ in atomic units [13]. However, upon being accelerated further by the laser field, the wavepacket's center of mass motion quickly approaches the trajectory from the above classical treatment [24]. Because of this, the longitudinal velocity distribution is neglected in the remainder of this thesis.

Ellipticity dependence

In this thesis ion yields for different laser pulse ellipticities are compared. Introducing an ellipticity reduces the electric field amplitude, but also changes its time dependence. This has an impact on the average ionisation rate per laser pulse. The TI rate in Eq. 2.6 is presented in the quasistatic limit and is thus time-independent. In an experiment however, the total ion yield from an entire laser pulse is measured. This requires integrating $\Gamma^{(TI)}(E_L(t))$ over the duration of the pulse, which results in different average ionisation rates per pulse depending on its polarisation state. This becomes clear when comparing the electric field amplitude of a linearly and circularly polarised wave as limiting cases. An arbitrarily polarised electromagnetic wave can be written as:

$$\mathbf{E}(t) = E_x \sin(\omega_L t)\hat{x} + E_y \sin(\omega_L t + \phi)\hat{y} \tag{2.8}$$

Setting $E_x = E_y = E_L$, the wave is linearly polarised for $\phi = 0$ and circularly for $\phi = \pi/2$, such that:

$$|E^{(lin)}(t)| = E_L\sqrt{2}|\sin(\omega_L t)| \tag{2.9}$$

$$|E^{(cir)}(t)| = E_L \tag{2.10}$$

In the long pulse regime, it turns out that the ionisation yield is reduced when changing the laser pulse polarisation from linear to circular. Hence, for obtaining the same ionisation rate, the intensity and thus electric field of the circularly polarised pulse has to be increased. Theoretical calculations of the TI rate lead to $I^{(lin)} = 0.65I^{(cir)}$, as the intensity relationship for achieving this [25]. This has been verified experimentally for atoms and small molecules [25, 26], including CO_2 [27].

Multielectron effects

One particular shortcoming relevant in the context of this experiment is the single active electron (SAE) approximation, which neglects the states of the remaining electrons during the ionisation process. It has therefore been suggested that the rapid removal of an electron through ionisation may lead to non-adiabatic multielectron (NME) dynamics in the ion, resulting in core excitation [28, 29]. Such channels have been observed in atoms [30–32], diatomics [33] and large polyatomics [29] and predicted to be relevant in CO_2 [34]. Their dependence on the laser parameters, however, is not trivial as it involves the coupling of the contributing tunnel-ionisation channels with the time-evolution of the remaining electron states, including the electric dipole field created by the tunneling electron [33]. This will be discussed in more detail in the case of strong field ionisation in molecules (see Sect. 2.3.2).

Over-the-barrier ionisiation (OTBI)

When the laser intensity becomes high enough, the barrier can be suppressed below the ground state. This is illustrated in part (c) of Fig. 2.1. The electron can escape over the barrier without tunneling and the ground state is depleted. For this, the onset of OBTI is associated with ionisation saturation in the pure strong field regime. The corresponding critical intensity can be approximated as [3]:

$$I^{(OTBI)} = \frac{c_0 \epsilon_0^3 \pi^2 I_p^4}{2Z^2 e^6} \tag{2.11}$$

where Z is the charge state of the corresponding atom or ion.

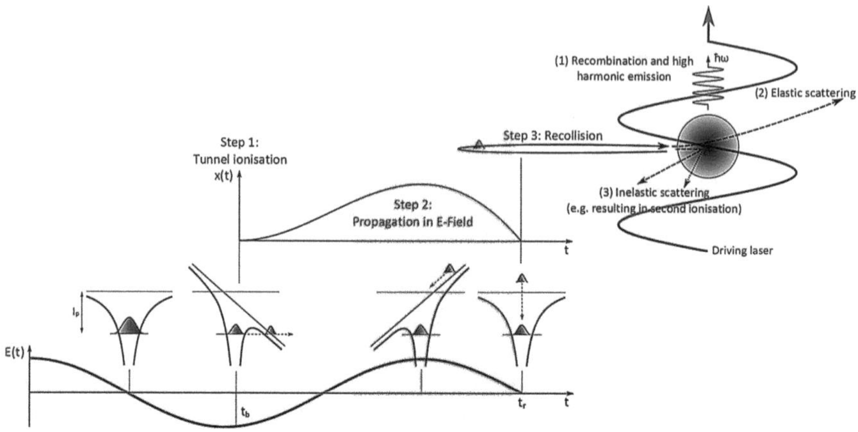

Fig. 2.3 Illustration of the 3-step model as explained in the text. The possible recollision effects are depicted in the inset at the *upper right corner* (Adapted from [38], with permission from Springer Science+Business Media)

2.1.3 Free Electron Dynamics and Recollison

After TI, the electron is rapidly driven away from the ion. However, due to the oscillatory motion of the laser field, the free electron can be driven back to recollide with the parent ion. Upon recollision the electron may recombine to its original ground state or scatter off the ion elastically or inelastically. Recombination results in high harmonic generation (HHG) (see Sect. 2.1.5), elastic scattering in high order ATI [35] and inelastic scattering may lead to further excitation or ionisation of the parent ion (see Sect. 2.1.4). Via this set of effects, laser driven electron recollision has established itself as a versatile tool to probe ultrafast molecular dynamics [36]. The first model for recollision was developed by Corkum [22] and Kulander et al. [37] and is based on a semiclassical treatment and called the 3-step model. It is illustrated in Fig. 2.3. Here, the first step is TI, followed by a classical propagation of the free electron as the second step. The third step is recollision with the possible outcomes mentioned above. However, a fully quantum mechanical treatment based on the SFA was developed soon after by Lewenstein et al. in the context of HHG [16].

Classical free electron propagation

The simplest approach for describing the propagation of the electron wavepacket is to treat its center of mass motion classically. Furthermore the SFA is applied by neglecting the influence of the Coulomb potential during the propagation in the continuum. This is justified as the electron is driven far away from the ion before the laser field approaches a zero crossing of its cycle. Here the electron trajectories are determined using a monochromatic, linearly polarized laser field:

$$\mathbf{E}_L(t) = -\frac{\partial \mathbf{A}(t)}{\partial t} = \mathbf{e}_z E_L \sin(\omega t) \tag{2.12}$$

where $\mathbf{e}_z$ is the unit vector along the z-coordinate axis. Equation (2.12) also defines the *vector potential* **A** of the electric field. The classical initial conditions for the tunneled electron have been presented in Sect. 2.1.2. Here they are written as $v(t_b) = 0$ and $z(t_b) = 0$, where t_b denotes the birth time of the electron within the laser cycle and $v = v_\parallel$ its longitudinal velocity along the z-axis. The motion of the electron is then determined by Newton's equations of motion, where I mainly follow the derivation of Lein [36]:

$$\ddot{z}(t) = -e/m_e E_L \sin(\omega t) \tag{2.13}$$

$$\dot{z}(t) = v(t) = \frac{eE_L}{m_e\omega}\Big(A(t) - A(t_b)\Big) = \frac{eE_L}{m_e\omega}\Big(\cos(\omega t) - \cos(\omega t_b)\Big) \tag{2.14}$$

$$z(t) = \frac{eE_L}{m_e\omega^2}\Big(\sin(\omega t) - \sin(\omega t_0) - \omega(t - t_b)\cos(\omega t_b)\Big) \tag{2.15}$$

From Eq. (2.14) one can infer that the electron velocity consists of a constant drift term $v_{drift} = -\frac{eE_L}{m_e\omega}\cos(\omega t_b)$ and an oscillating term. The resulting kinetic energy is then given by [36]:

$$E_{kin}(t, t_b) = \frac{1}{2}m_e v^2(t) = 2U_p\Big(\cos^2(\omega t) - 2\cos(\omega t)\cos(\omega t_b) + \cos^2(\omega t_b)\Big) \tag{2.16}$$

Recollision

The condition for recollision in a linearly polarised field is the return of the electron to the vicinity of the parent ion, before it is driven away again in the next laser cycle:

$$z(t_r) \leq \sigma_{ion} \tag{2.17}$$

where σ_{ion} is the recollision cross-section of the parent ion; typically assumed to be on the order of the Bohr radius a_0, such that $\sigma_{ion} \approx a_0^2 \approx 10^{-21}$ m^2, depending on the size of the atom or molecule. Here, t_r is the recollision time with $t_r > t_b$ such that $\tau = t_r - t_b$ is the *return time* of the recolliding electron. Whether recollision takes place or not only depends on the initial conditions and thus birth phase ωt_b of the free electron. Via Eq. 2.16, the recollision energy can then be calculated as $E_r = E_{kin}(t_r, t_b)$. It was first shown by Corkum [22] that the maximum kinetic energy upon recollision can be approximated as:

$$E_r^{(max)} \approx 3.17 U_p \propto I\lambda^2 \tag{2.18}$$

corresponding approximately to a birth phase of 17° and a recollision phase of 270° after a peak of the electric field [39]. Note that the recollision energy scales linearly with intensity and quadratically with the laser wavelength. A longer wavelength leads to more time spent in the continuum, allowing the electron to be accelerated for a longer time.

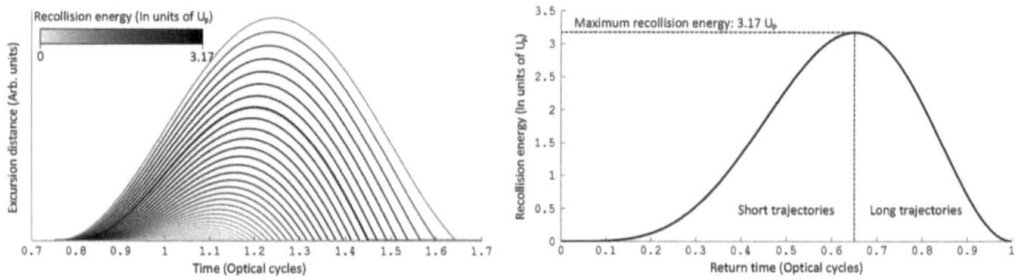

Fig. 2.4 Classical recollision trajectories from the second peak in a sinusoidal laser cycle; calculated via the formulae presented in the text. Both panels illustrate that for each recollision energy there are two types of trajectories: one with a short and one with a longer return time (Reprinted from [38], with permission from Springer Science+Business Media)

A set of trajectories with identical birth phases is launched every half cycle of the laser field, but in opposite directions. However, each trajectory within such a set occurs with a probability given by the TI rate resulting from the associated birth phase within the laser cycle. The exponential decay of the TI rate with the laser electric field thus acts as a narrow filter that confines the contributing birth phases to the peak region of each half cycle. A set of classical trajectories from a sinusoidal laser half-cycle is displayed in Fig. 2.4.

Note that any momentum component perpendicular to the tunneling direction of the electron will drive it away from the remaining ion. This manifests itself in a strong laser polarisation dependence of the recollision step, such that the electron cannot recollide for circular polarisation. In the case of rare gas atoms and small molecules it has in fact been observed experimentally that the electron does not recollide for laser ellipticities larger than $\epsilon \approx 0.3$ [40].

However, there are exceptions to this rule. First, the transverse velocity spread of the electron wavepacket (see Eq. 2.7) may drive the electron away from the ion, reducing the recollision probability. An elliptically polarised laser pulse can then compensate for the initial transverse momentum. This means that recollision may take place at higher ellipticities. In the case of atoms, this effect has thus far only been suggested to be significant in Mg [41]. In the case of molecules, their orbital structure may increase the transverse velocity spread of the tunneled electron. This can then lead to a maximum recollision probability for non-zero ellipticity. This effect is discussed in detail in Chap. 6. Second, when the electron gets close to the parent ion upon its return, it may be attracted by its Coulomb potential. This *Coulomb focussing* effect may compensate for a transverse offset of the recolliding electron in an elliptical driving laser field [42]. Here, it should also be mentioned that the ejected electron may only recollide after several laser cycles (see for example the analysis of [43]).

Direct electrons and elastic scattering

If the electron does not recollide and the laser pulse is longer than a few cycles, the electric field oscillations acting on the electron will vanish adiabatically. Therefore, the kinetic energy and momentum of such a *direct electron* is determined by its drift velocity and hence birth phase:

$$E_{drift} = 2U_p \cos^2(\omega t_b) \quad (2.19)$$

$$p_{drift} = 2\sqrt{U_p}\cos(\omega t_b) \quad (2.20)$$

Note that the maximum drift energy $E_{drift}^{(max)} = 2U_p$ is acquired at the zero crossing of the electric field (peak of the vector potential). If ionisation takes place at the peak of the electric field (zero crossing of the vector potential), the corresponding drift energy will be zero. The same holds true for the drift momentum. Direct electrons are responsible for the plateau observed in ATI [35]. For completeness it should also be mentioned that if the electron recollides elastically it will gain drift energy in addition to the recollision energy. Numerical calculations based on the classical propagation show that a maximum drift energy of about $10U_p$ can be reached in this case [35]. This energy corresponds to the cut-off in high order ATI spectra in the tunneling regime.

Quantum mechanical picture and further corrections

In the quantum mechanical picture, the response of an atom to a laser field is modeled by the time-dependent Schrödinger equation (TDSE). Employing the single active electron and dipole approximation the TDSE takes the following form in the so-called length gauge (see for example [6]):

$$i\hbar\frac{\partial}{\partial t}\psi(\mathbf{r},t) = \left[-\frac{\hbar^2\nabla^2}{2m} + V(\mathbf{r}) + \boldsymbol{\mu}\cdot\mathbf{E}_L(t)\right]\psi(\mathbf{r},t) \quad (2.21)$$

where $V(\mathbf{r})$ is a one-electron potential and $\boldsymbol{\mu}$ the dipole transition operator that describes the electronic response to the laser field $\mathbf{E}_L(t)$. The above TDSE is the starting point for many theoretical treatments including the SFA and of course ab initio calculations. As a more detailed discussion is beyond the scope of this thesis, I will only highlight the most significant modifications of the quantum mechanical picture with respect to the previously discussed classical model that are relevant to this thesis. Here, the recolliding electron is a fraction of the ground state population driven into the continuum. Recollision can then be interpreted as the interaction between the remaining ground state population ψ_g and the returning continuum wavepacket ψ_c. The evolution of the continuum wavepacket is often modeled via the Feynman path integral approach. Here, the phase accumulated by the spatial parts of the wavepacket over time is related to the action S along the classical trajectory presented previously [12, 24]:

$$\psi_c(\mathbf{r},t) \propto \sqrt{\Gamma^{(TI)}(t_b)}\exp\left[iS(\mathbf{r},t,t_b)\right] \quad (2.22)$$

where the TI rate determines the initial wavepacket distribution, as discussed earlier. Generally, different spatial parts of the wavepacket correspond to different momenta along the classical trajectory and thus different energies. Within the SFA framework, the Coulomb potential is neglected for the propagation and the wavefunction of the unperturbed free electron becomes a plane wave, called *Volkov state*. Here, an analytical expression for the action can be obtained [(in atomic units (a.u.)]:

$$S(\mathbf{r}, t, t_b) = [\mathbf{A}(t) - \mathbf{A}(t_b)]\mathbf{r} - \frac{1}{2}\int_{t_b}^{t} [\mathbf{A}(t') - \mathbf{A}(t_b)]\, dt' \tag{2.23}$$

where $\mathbf{v}(t) = \mathbf{A}(t) - \mathbf{A}(t_b)$ is electron's instantaneous kinetic momentum along the classical trajectory.

In this picture, the recollision probability depends on the overlap between the returning continuum wavepacket and the remaining ground state population. Two properties of the electron wavepacket may reduce the probability: wavepacket spreading in the continuum and its transverse velocity spread upon its birth. Wavepacket spreading increases with the time spent in the continuum. This only depends on the driving laser wavelength λ. This effect has been investigated theoretically with HHG as a probe. Here, numerical solutions to the time-dependent Schrödinger equation (TDSE) for the HHG emission of a single atom have shown a reduction of the emission scaling as $\lambda^{-5.5\pm0.5}$ [44].

2.1.4 Nonsequential Double Ionisation

Non-sequential double ionisation describes the double ionisation of an atom or molecule, for which the two ionisation steps are correlated. This is not the case for *sequential double ionisation*, where each ionisation happens independently of the other. NSDI was first observed in form of a knee structure in the ion yields in Xenon as a function of intensity [45, 46]. Along with the recollision model, where the two ionisation steps are separable in time, two instantaneous mechanisms were proposed to explain the effect: *shake-off* [47] and *collective tunneling* [48]. In shake-off, a single high energy photon is absorbed ($\hbar\omega \gg I_p$), such that the ionisation process is so fast that the remaining electrons cannot adiabatically readjust to form the orbital of the ion [49]. The resulting non-adiabatic multi-electron dynamics then cause excitations that lead to further ionisation of the parent ion. In collective tunneling, multiple electrons simultaneously tunnel through the finite barrier. In its first proposal, however, Eichmann already points out that ionisation rates of this mechanism are quantitatively too low to account for laser driven NSDI. Fittinghoff et al. [50] then give a first strong indication for rescattering being the main mechanism, by measuring the polarisation dependence of He^{++} and Ne^{++} yields. In particular a suppression of the knee structure for circular polarisation was observed, which can only be explained in the rescattering model.

Within the recollision model, two NSDI channels are possible (see for example [51]):

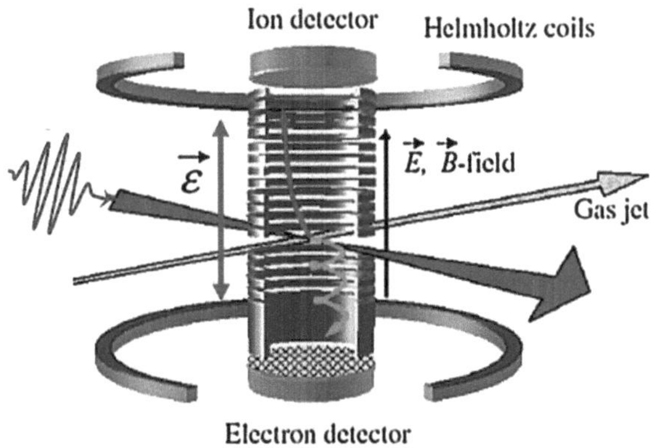

Fig. 2.5 Schematic sketch of the reaction microscope typically used for differential recoil ion measurements. The sample is delivered in form of a rotationally cold supersonic gasjet, perpendicular to the laser propagation direction. Ions and electrons are extracted via electric and magnetic fields respectively and detected in coincidence via position-sensitive micro-channel plate detectors (MCP) in conjunction with delay-line anodes (Reprinted from [51]. ©IOP Publishing and Deutsche Physikalische Gesellschaft. CC BY-NC-SA)

1. *Recollision induced impact ionisation* takes place when the kinetic energy of the recolliding electron is higher than the ionisation potential of the remaining ion I_p^+, such that the second electron is released instantaneously upon recollision. Note that in this case, the recollision phase of the first electron corresponds to the birth phase of the second one. This channel is sometimes denoted as (e, 2e).
2. *Recollision-excitation with subsequent field ionisation* (RESI) was first proposed by Feuerstein et al. [52] and takes place when the kinetic energy of the electron is below I_p^+. In this case the ion is lifted to an excited state upon recollision, such that it can be field-ionised near one of the subsequent peaks of the laser field. Note that there is a time delay between recollision of the first and release of the second electron.

In order to provide clear evidence for the recollision model and a separation of the two possible channels, numerous differential measurements of the recoil ion momentum in NSDI were performed [51–55]. Using a reaction microscope like in *Cold Target Recoil Ion Momentum Spectroscopy* (COLTRIMS, for a review see for example the article by Dörner et al. [56], a simplified setup is illustrated in Fig. 2.5), the recoil momentum vectors of the doubly ionised ion and the two ejected electrons are determined via their position on the detector and time-of-flight (TOF). A kinematic analysis of these vectors is based on momentum conservation during recollision (neglecting photon momenta) and the fact that the two electrons are detected as directed electrons, with drift momenta corresponding to Eq. (2.19). Note that in this case the relevant phase of the recolliding electron is its recollision phase, as this is the point in time when it is eventually driven into the continuum by the laser field. The electron and ion momenta therefore reveal the laser field phase in which the particles were released and therefore enable one to distinguish between the NSDI channels described above. This is illustrated and summarised in Fig. 2.6. Several studies have

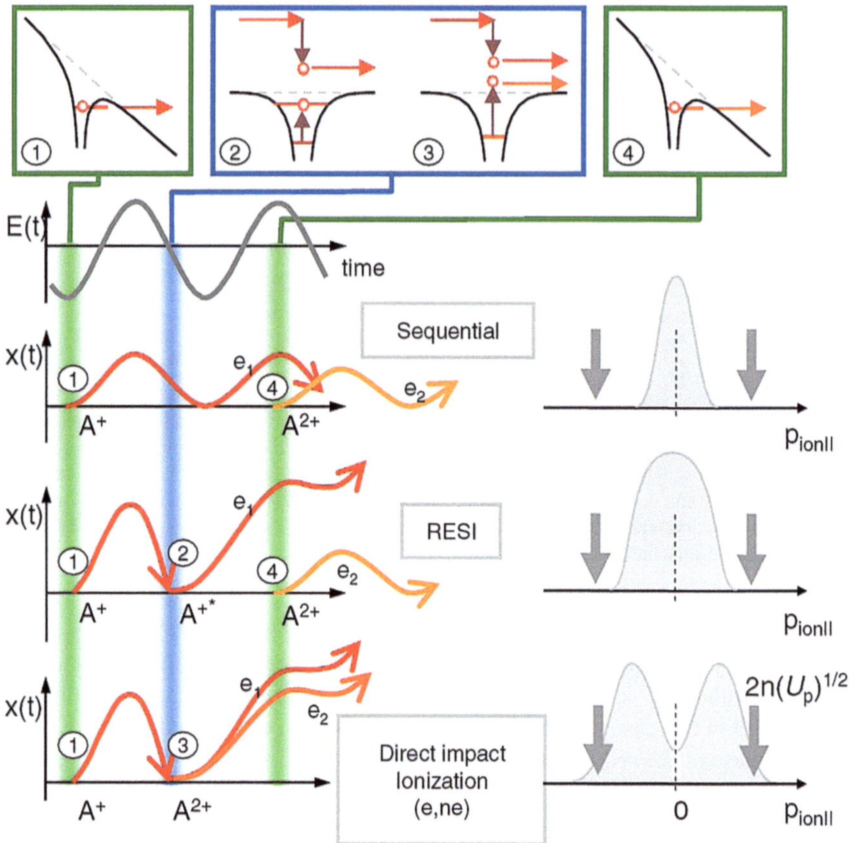

Fig. 2.6 Summary of the possible double ionisation channels (taken from [57]). The first electron is created at (*1*) via tunneling ionisation. In sequential ionisation, the second electron is tunnel ionised after one cycle at the peak of the field (*4*) and hence both electrons have very small momenta. This is shown in the corresponding ion momentum distribution. In RESI, the first electron recollides at (*2*), excites the atom and a second electron is tunnel ionised afterwards at (*4*). This leads to a broader ion momentum distribution around zero due to the excitation energy. In direct impact ionisation, both electrons acquire a large drift momentum, such that the ion momentum distribution has a characteristic dip at zero. The *arrows* point at the maximum drift momenta (Reprinted with permission from [57]. Copyright (2004) by the American Physical Society)

been conducted to determine the dependence on atomic structure and laser pulse parameters of the three double ionisation channels. They are briefly reviewed in the following paragraphs.

Atomic structure

The dependence on the atomic structure can be illustrated by comparing a light atom like Ne to a heavier one like Ar (see for example [51, 58, 59]). For comparable intensities, the RESI channel is much more significant in Ar than in Ne, as shown

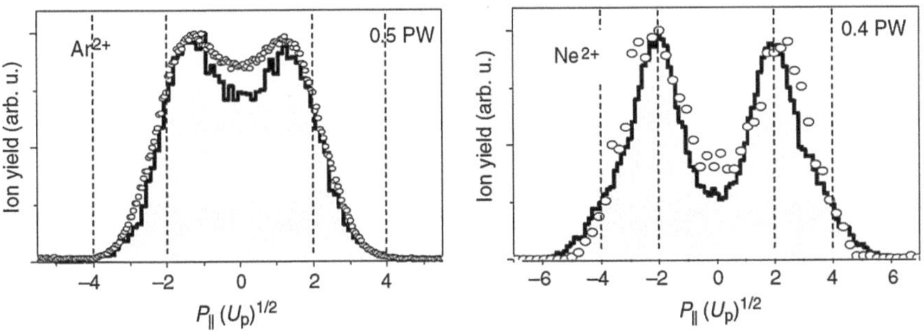

Fig. 2.7 Ion momentum spectra from Ar and Ne at 800 nm (*circles*) and 1300 nm (*solid line*) [51]. The experiment was performed using 30 fs at 800 nm and 35–40 fs at 1300 nm pulses with the intensity denoted in PWcm^{-2} in the figure. The momenta are given in units of $\sqrt{U_p}$ in order to account for the difference in ponderomotive potential (Reprinted from [51]. © IOP Publishing and Deutsche Physikalische Gesellschaft. CC BY-NC-SA)

in Fig. 2.7. By comparing ion momentum spectra with similar ratios of recollision energy to the second ionisation potential $3.17U_p/I_p{}^+$, de Jesus et al. [58] attribute this to the larger excitation cross-section of Ar. Alnaser et al. [59] argue that the scaling of the above recollision energy ratios is limited by the atom's saturation intensity I_{sat}. Therefore the difference in ionisation potentials cannot be excluded at all intensities in this way but the authors argue that it will always manifest itself in the Keldysh adiabaticity parameter $\gamma = \sqrt{I_p/U_p}$. Thus at 800 nm $\gamma^{(Ar)} = 0.6$ and $\gamma^{(Ne)} = 0.4$, such that Ne is further in the tunneling regime and hence favouring the recollision scenario with direct impact ionisation.

Laser pulse parameters

Broad studies have been performed by [55, 59] for example. Starting from a direct impact ionisation scenario, both decreasing and increasing the intensity will lead to a single peak ion momentum distribution [55]. For low intensities the RESI channel will dominate, as the recollision energy drops below $I_p{}^+$ and the first ionisation step will enter the multi-photon regime. For high intensities the sequential ionisation channel will dominate. The wavelength dependence is qualitatively similar to intensity, but quantitatively stronger due to its quadratic scaling in U_p. Several studies [51, 59] show that the impact ionisation channel becomes increasingly dominant for higher wavelengths. This is due to an increase in recollision energy and decrease in the Keldysh parameter and hence going deeper into the tunneling regime. Figure 2.7 shows that for longer wavelengths the dip at zero momentum becomes more pronounced, which corresponds to a stronger impact ionisation channel.

In addition to the above discussion, Liu et al. [60] showed that the carrier envelope phase (CEP) of a few-cycle pulse influences the NSDI ion momentum spectra, by introducing a pronounced asymmetry in the ion ejection direction. The experiment was performed in Ar using 5 fs, 500 μ J pulses at 760 nm. Nevertheless, their numerical simulation showed that this impact vanishes for pulses as long as 8 cycles, which

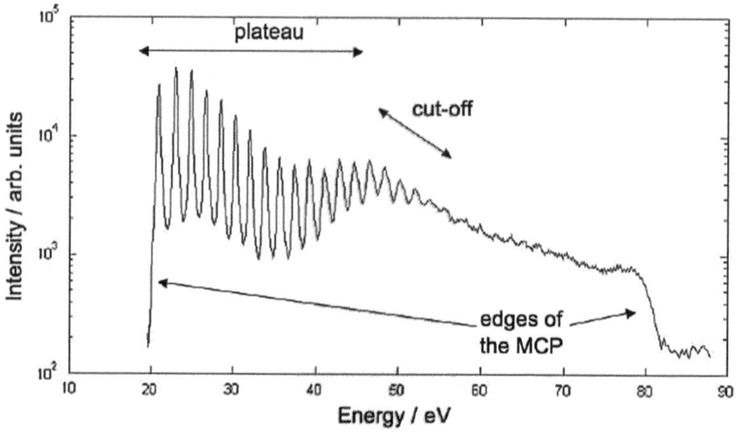

Fig. 2.8 Typical HHG spectrum that displays the essential features described in the text. Here, MCP refers to the multichannel plate detector used to record the spectrum (Figure and measurement courtesy of Thomas Siegel)

corresponds to approximately 21 fs at 800 nm. This effect is thus not relevant in the context of this thesis.

2.1.5 *High Harmonic Generation*

Upon recollision, the electron may recombine with the parent ion. This is followed by the emission of its recollision energy in form of a single photon, which is an integer multiple of the driving laser photon energy. The process is thus called High Harmonic Generation (HHG). In the quantum mechanical treatment, the recombination of the free electron with the ground state is modeled as rapid oscillations of the resulting orbital wavefunction and hence atomic electron density. The recollision process thus represents a time-dependent induced dipole moment $\mu_h(t)$, which causes the emission of high harmonic radiation with intensity I_h due to its acceleration [12]:

$$I_h(t) \propto |\ddot{\mu}_h(t)|^2 \tag{2.24}$$

High harmonic emission takes place twice every laser cycle. Due to the coherence of the driving laser field, the harmonic emission of the individual half cycles interfere and lead to an HHG spectrum with discrete peaks spaced by twice the driving laser frequency (see Fig. 2.8). In a monoatomic gas the emission obeys inversion symmetry, such that $\mu_h(t + T_L/2) = -\mu_h(t)$. This implies that the harmonic spectrum only consists of odd integer multiples of the driving laser frequency [3].

In the multiphoton regime, the harmonic intensity decreases very quickly with the harmonic order. This can be understood from the MPI rate from (2.1), as it decreases rapidly with the number of photons involved. For higher intensity, one enters the tunneling regime and the observed harmonic spectra change. After the decrease of harmonic intensities for the first orders, the spectrum shows a plateau in intensity, which is followed by a very quick cut-off. The cut-off is due to the maximal kinetic energy, the electron can acquire in the driving laser field. Therefore, one can extend the cut-off, by either increasing the intensity or the laser wavelength λ. Both approaches have limitations. The intensity cannot be increased infinitely as one will eventually reach the saturation intensity for the ionisation step. For this reason it is often advantageous to use atoms or molecules with a high I_p in order to increase the laser intensity as much as possible. The wavelength seems to be the better option as the ponderomotive energy scales quadratically with it. Yet the higher wavelength also leads to a rapid increase of wavepacket spreading as discussed earlier. These effects have to be balanced in order to generate high harmonics efficiently.

HHG has become a powerful technique for studying molecular dynamics. On the one hand, one can use the emission spectrum itself to investigate the molecular properties of its source. Due to the return time to recollision energy mapping for each trajectory, the HHG spectrum contains information about the parent ion dynamics during the electron's excursion time (see for example [61]). On the other hand the properties of the emission (ultrahort pulse duration, high brightness, good coherence [6]) open up the possibility of using HHG as a laser pulse source for producing pulses with a duration of a few hundred attoseconds [12]. This in turn provides the opportunity do develop pump-probe like experiments for investigating electron dynamics in atoms and molecules on this time scale.

2.2 Molecules

Accurate modelling of molecular dynamics is an ongoing challenge in physics due to the large number of interacting particles. This section aims at presenting the most common approximations for a qualitative understanding of the dynamics studied in this thesis. They will be restricted to the case of diatomics in time-independent fields. Given that CO_2 is a symmetric, linear molecule, this is a reasonable approach. Extensions due to the possible bend of the molecule and its detailed ionisation and dissociation dynamics are discussed in the respective later chapters.

2.2.1 Electronic Structure

On first sight, molecules might not be very different from atoms: it still comes down to solving the time-independent Schrödinger equation (TISE). The solutions are the molecular eigenstates $\Psi_M(\mathbf{r}, \mathbf{R})$, each uniquely associated with an energy eigenvalue E_M. The latter are the energy levels observed in experiments and are thus important for understanding the evolution of the molecular system. The Born-Oppenheimer approximation (BOA) leads to a rather accessible description by decoupling electronic and nuclear energy within the molecule. In strong field physics however, the electron density profile or *orbital* is relevant for understanding ionisation in the molecular frame of reference. It is connected to the molecular eigenstate within the framework of the molecular orbital theory. The energy level structure and orbital picture really are two sides of the same coin and are conveniently connected via the classification of quantum states in spectroscopic notation.

The Born-Oppenheimer approximation

The general molecular Hamiltonian can be written in the following form (see for example [62, 63]):

$$H_M = \underbrace{\sum_i -\frac{\hbar^2\nabla_{e,i}^2}{2m}}_{\text{electron kinetic energy}} + \underbrace{\sum_{j>i} \frac{e^2}{|\mathbf{r}_i - \mathbf{r}_j|}}_{\text{electron-electron repulsions}} + \underbrace{\sum_i -\frac{\hbar^2\nabla_{N,i}^2}{2M_i}}_{\text{nuclear kinetic energy}} + \underbrace{\sum_{j>i} \frac{Z_iZ_je^2}{|\mathbf{R}_i - \mathbf{R}_j|}}_{\text{nuclear-nuclear repulsions}}$$
$$+ \underbrace{\sum_{ij} -\frac{Z_je^2}{|\mathbf{r}_i - \mathbf{R}_j|}}_{\text{electron-nuclear attractions}} \equiv T_e + V_e + T_N + V_N + V_{e,N} \tag{2.25}$$

where the $\{\mathbf{r}, \nabla_e\}$ refer to the coordinates and momenta of the electrons and $\{\mathbf{R}, \nabla_N\}$ to the corresponding properties of the nuclei. Z_i denotes the nuclear charge of nucleus i. Apart from the electronic configuration, a molecule has additional degrees of freedom and thus associated energy levels, because of the relative motion of the nuclei. These are vibrations along the molecular axis and rotations around the molecule's center of mass. The BOA now assumes that the electronic configuration can be decoupled from the nuclear motion. This is justified by the large difference between electronic and nuclear mass, such that the electrons move much faster than the nuclei. Therefore, the nuclear geometry is described as a quasistatic framework for the electrons. Their wavefunction then depends only parametrically on the nuclear coordinates $\{\mathbf{R}\}$. The molecular wavefunction can then be separated into an electronic $\psi_e(\mathbf{r}, \mathbf{R})$ and a nuclear part $\psi_N(\mathbf{R})$:

$$\Psi_M(\mathbf{r}, \mathbf{R}) = \psi_e(\mathbf{r}, \mathbf{R})\psi_N(\mathbf{R}) \tag{2.26}$$

This implies that electronic, vibrational and rotational contributions to the molecular energy level structure can also be separated [64]:

$$E_M = E_e + E_{vib} + E_{rot} \tag{2.27}$$

With the internuclear geometry fixed, the electronic wavefunction must fulfill the time independent Schrödinger equation (TISE):

$$H_e\psi_e(\mathbf{r}, \mathbf{R}) = (T_e + V_e + V_{e,N})\psi_e(\mathbf{r}, \mathbf{R}) = E_e(\mathbf{R})\psi_e(\mathbf{r}, \mathbf{R}) \tag{2.28}$$

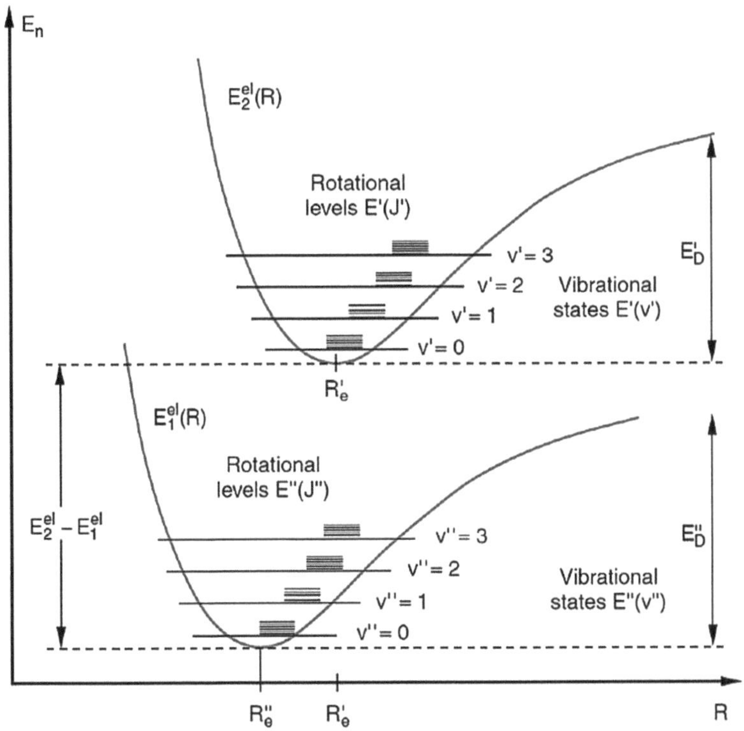

Fig. 2.9 Schematic representation of the rovibrational energy level structure of two electronic states of a diatomic molecule. Here, R_e denotes the equilibrium bond length and E_D the dissociation energy (Reprinted from [65], with permission from Springer Science+Business Media)

Plugging (2.26) and (2.28) into the TISE for the molecular Hamiltonian and neglecting the change of the electronic wavefunction with respect to the nuclear coordinates, one obtains [62, 64]:

$$H_M \psi_N(\mathbf{R}) = (T_N + E_e(\mathbf{R}) + V_N(\mathbf{R}))\psi_N(\mathbf{R}) = E_M \psi_N(\mathbf{R}) \qquad (2.29)$$

The molecular energy eigenstate E_M was thus obtained from the nuclear motion in an effective potential consisting of the fast and thus averaged electron-electron and electron-nuclear interaction contained in $E_e(\mathbf{R})$ and the instantaneous internuclear interaction contained in $V_N(\mathbf{R})$. This effective potential can be calculated for every internuclear geometry by solving the electronic TISE (2.28) and including $V_N(\mathbf{R})$, hence giving the potential energy surface (PES) on which the nuclei move [63]. The energy level structure associated with the nuclear part is contained in the molecular energy eigenvalue via Eq. 2.27. In the BOA framework it is thus a fine structure superimposed onto the PES associated with the electronic energy. This is illustrated

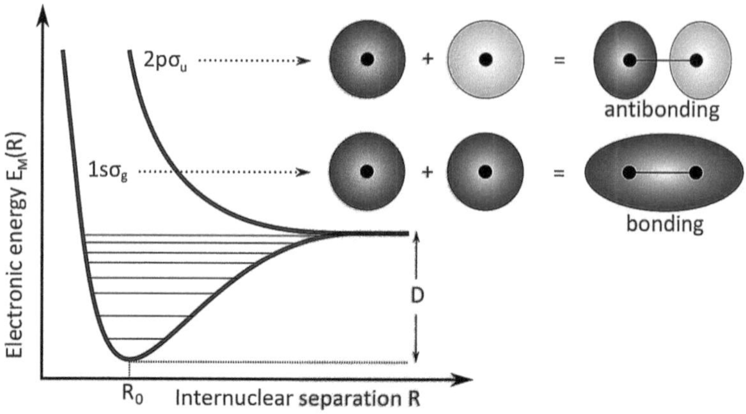

Fig. 2.10 Schematic representation of the PECs corresponding to the bonding and antibonding state of H_2^+ including a graphical representation of the associated linear combination of atomic orbitals as explained in the text. Here, the different shadings of the orbitals correspond to different signs of the electronic wave function

in Fig. 2.9. Note that for a diatomic molecule, the PES is reduced to a potential energy curve (PEC).

Molecular orbital theory

The PECs obtained within the BOA framework illustrate the requirement for chemical bond formation. A local minimum means that the potential energy of the nuclei is lowest for an associated equilibrium bond length R_0. The energy D that is required to overcome the potential well, such that $R \to \infty$, is called the *dissociation energy* or *limit* of this bound state. This is illustrated in Fig. 2.10. Note that this energy is slightly larger than the real chemical dissociation energy, as the vibrational ground state of the molecule is not taken into account [66]. A molecular orbital is proportional to the electronic probability density $|\psi_e(\mathbf{r}, \mathbf{R})|^2$ within the nuclear framework. The molecular orbital corresponding to a bound state typically has a non-zero probability density between the atoms. In principle a calculation of the bound states and their orbitals via the BOA would suffice for a complete description of a molecule. However, exact solutions can only be obtained for the simplest possible molecule H_2^+. In order to simplify the calculations for more complex systems, one may assume that only the electrons on the outer shells contribute to the bond formation, whereas the ones close to the nucleus remain unperturbed [67]. This implies that for distances close to one of the nuclei, the molecular orbital converges to the corresponding atomic one. As the TISE is linear, this suggests that molecular orbitals may be approximated as linear combinations of atomic orbitals (LCAO).

Formally, one would assume that the atomic wavefunctions $\{\psi_{A,i}\}$ form a basis of the molecular Hamiltonian and then take their linear combination as the trial solution for the molecular wavefunction [68]:

$$\Psi_{LCAO} = \sum_i c_i \psi_{A,i} \tag{2.30}$$

For the hydrogen molecular ion, this results in the wavefunctions $\Psi_{LCAO,+} = \psi_{H,1} + \psi_{H,2}$ and $\Psi_{LCAO,-} = \psi_{H,1} - \psi_{H,2}$. They are shown graphically as an inset in part (a) of Fig. 2.10. One observes that for the symmetric linear combination, the resulting wavefunction has a non-zero probability density between the nuclei and is thus called a *bonding* orbital. For the anti-symmetric combination, the density drops to zero there and an *antibonding* orbital is formed. This is directly reflected in the associated PECs, corresponding to a bound and repulsive state, respectively.

Classification of quantum states

As was mentioned previously, every electronic energy eigenstate is associated with a molecular orbital. Each orbital can be uniquely classified by its symmetry properties, which is widely used in spectroscopy to label the energy levels. For diatomics, the three symmetry properties are: rotational symmetry around the molecular axis, inversion symmetry with respect to the molecular centre and reflection symmetry at a plane containing the molecular axis. Formally, the mathematical operators that perform these transformations leave the molecular Hamiltonian invariant and commute, such that they span a complete set of simultaneous eigenfunctions of the molecular system. This means that their quantum numbers uniquely identify each possible electronic energy level [68].

The rotational symmetry is associated with the conservation of the molecule's angular momentum component L_z along the molecular axis z. The corresponding eigenvalue equation with the quantum number Λ is given by [69]:

$$L_z \psi_e = \pm \Lambda \hbar \psi_e, \quad \Lambda = 0, 1, 2, \ldots \tag{2.31}$$

The labeling of the states is done with greek letters as in atoms, such that $\Lambda = 0, 1, 2, \cdots \leftrightarrow \Sigma, \Pi, \Delta, \ldots$. Inversion symmetry is associated with the notion of *parity* of the electronic wavefunction. If the sign of the wavefunction changes (does not change) upon inversion its parity is odd (even). In spectroscopic notation the subscript u (g) is used. For reflection symmetry a change in sign is denoted by a '$-$' superscript. Invariance under reflection is then denoted by a '$+$' superscript. In orbitals with $\Lambda > 0$, this concept is meaningless due to the broken cylindrical symmetry of the wavefunction. This classification is therefore only relevant for Σ states. Note that for atomic orbitals the analogue notation in lower-case greek letters is used for the projection of the angular momentum quantum number. This notation is often used when referring to single electron states within a molecular system [69].

Lastly, a particular quantum state may be degenerate due to the spin of its unpaired electrons. If all electrons are paired, the state is 1-fold degenerate (a singlet). If there is one unpaired electron it can have spin up or down, so the state is 2-fold degenerate (doublet), whereas two unpaired electrons result in a triplet state. This is denoted as the multiplicity $2S + 1$ via the total spin S of the molecular eigenstate. The full term symbol of the state is now written as ${}^{2S+1}\Lambda^{+/-}_{g/u}$. In addition to the term symbol, energy levels in a molecule are usually ordered according to their excitation energy, starting with the ground state: X, A, B, C, etc.

2.2.2 Radiative Transitions

When a molecule is placed in a laser field, its charge distribution interacts with the laser electric field. Here, the molecule may absorb energy and undergo a transition from its initial energy state Ψ_i to a final state Ψ_f. Within the dipole approximation (see Sect. 2.1.2) the corresponding light-molecule interaction can be treated as a dipole transition (higher multipole interactions are neglected). Neglecting the rotational energy and using the BOA, the state can be separated into electronic ψ and vibrational contributions ν, such that $\Psi = \psi\nu$. Here I have dropped the dependence on the electronic and nuclear coordinates for convenience. Analogously, the transition dipole moment $\boldsymbol{\mu}$ between the states consists of an electronic and a nuclear part with $\boldsymbol{\mu} = \boldsymbol{\mu}_e + \boldsymbol{\mu}_N$. In this way, the transition probability $M_{i\to j}$ can be obtained [69]:

$$\langle \Psi_f | \boldsymbol{\mu} | \Psi_i \rangle = \langle \psi_f \nu_f | \boldsymbol{\mu}_e + \boldsymbol{\mu}_N | \psi_i \nu_i \rangle \tag{2.32}$$

$$= \langle \psi_f | \psi_i \rangle \langle \nu_f | \boldsymbol{\mu}_N | \nu_i \rangle + \langle \nu_f | \nu_i \rangle \langle \psi_f | \boldsymbol{\mu}_e | \psi_i \rangle \tag{2.33}$$

If an electronic transition takes place, $\psi_i \neq \psi_f$ and due to their orthogonality $\langle \psi_f | \psi_i \rangle = 0$. Therefore, the transition probability turns out to be:

$$M_{i\to f} \propto |\langle \Psi_f | \boldsymbol{\mu} | \Psi_i \rangle|^2 \tag{2.34}$$

$$\propto |\langle \nu_f | \nu_i \rangle|^2 \times |\langle \psi_f | \boldsymbol{\mu}_e | \psi_i \rangle|^2 \tag{2.35}$$

The interpretation of the above equation is that the transition probability depends on the dipole transition probability between the two electronic states and on the overlap between initial and final vibrational states.

Electronic transitions

A dipole transition is *allowed* if $\langle \psi_f | \boldsymbol{\mu}_e | \psi_i \rangle \neq 0$. This is only possible if the total integrand of the corresponding integral does not have an odd parity [64]. The parity of the dipole transition operator $\boldsymbol{\mu}_e = -e\mathbf{r}$ is odd, because it has a direction. Therefore only transitions between states of opposite parity are allowed.

Depending on the direction of the transition dipole with respect to the molecular axis, the associated angular momentum Λ may be changed. Direct inspection of the integrand for dipole moments along and perpendicular to the molecular axis leads to allowed transitions for $\Delta\Lambda = 0$ and $\Delta\Lambda = \pm 1$, respectively [64]. This has an important consequence for transitions induced via linearly polarised light in the molecular frame. As the direction of the transition dipole moment is given by the laser field polarisation, this means that only the field component parallel to the molecular axis contributes to transitions with $\Delta\Lambda = 0$. The same reasoning applies

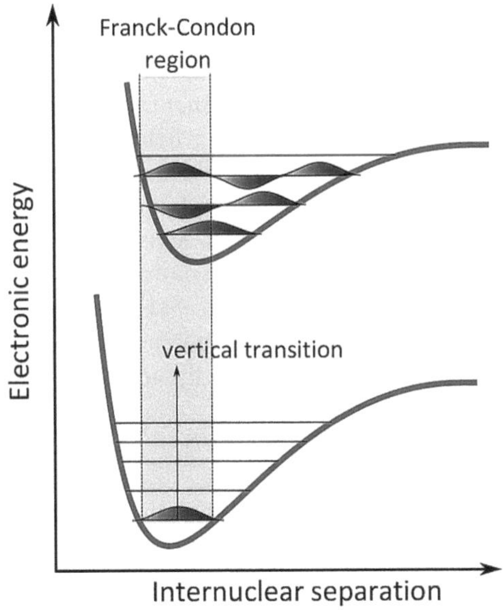

Fig. 2.11 Illustration of the Franck-Condon principle. Assuming that the transition takes place instantaneously, the initial vibrational wavepacket is projected vertically upwards. The transition probability to the final vibrational states scales with their overlap with the projection

to the perpendicular case. This implies the following polarisation dependences for transitions between $\Sigma \rightarrow \Sigma$ ($\Delta\Lambda = 0$) and $\Sigma \rightarrow \Pi$ ($\Delta\Lambda = \pm 1$) states:

$$M_{\Sigma\rightarrow\Sigma} \propto |\cos(\theta)|^2 \tag{2.36}$$

$$M_{\Sigma\rightarrow\Pi} \propto |\sin(\theta)|^2 \tag{2.37}$$

where θ is the angle between the polarisation direction and the molecular axis. As the transition dipole moment operator does not change the total spin S or the reflection symmetry in a parallel transition, the selection rules $\Delta S = 0$ (neglecting spin-orbit coupling) and $\Sigma^{\pm} \rightarrow \Sigma^{\pm}$ hold in addition to the above [64].

Vibrational transitions

By using the BOA it is implicitly assumed that the transition takes place in a quasi-static nuclear framework. This means that the transition is instantaneous without a change in the internuclear distance R. In the context of vibrational transitions, this approximation is called the *Franck-Condon principle* [70, 71]. As derived above, the transition probability to a vibrational state during an electronic transition is given by the overlap between the vertically projected initial vibrational state onto the possible final state. This is illustrated in Fig. 2.11. It is thus called a *vertical* transition, where the corresponding projection region is called the *Franck-Condon region*. The overlap integral $|\langle\nu_f|\nu_i\rangle|^2$ is called the *Franck-Condon factor*.

2.2.3 Field-Dressed States

In the Sect. 2.2.2, the energy level structure and related transitions have been treated perturbatively, meaning that the laser field does not couple to the molecular states. However, in the strong field regime this is not sufficient and the PECs can be significantly altered due to the presence of the field. For a non-perturbative approach, one would work with the time-dependent Schrödinger equation (TDSE) and include the laser field and its interaction with the molecule exactly. The total Hamiltonian of this system would thus consist of the time-independent field free molecular part H_M, the laser field in the quasistatic approximation H_L and the time-dependent interaction with the laser field $E_L(t)$ within the dipole approximation [69]:

$$H_{tot}(t) = H_M + H_L + \boldsymbol{\mu} \cdot \mathbf{E}_L(t) \tag{2.38}$$

A long laser pulse (>10 fs at 800 nm) may be assumed to consist of many identical cycles. Thus the dipole interaction with the molecule will also be approximately periodic with the laser frequency. As in the context of HHG (see Sect. 2.1.5), a periodically driven process in time is related to a spectrum of harmonics of the driving field via Fourier transformation [72]. In the energy domain, this corresponds to an infinite set of coherent states spaced by the photon energy $\hbar\omega_L$ of the laser field. This is the essence of the so-called *Floquet* theorem that thus leads to periodic eigenfunctions $\{\Psi_{tot,i}\}$ with infinite sets of eigenenergies $\{E_{tot,i}\} \pm n\hbar\omega$, where $n = 0, 1, 2, \ldots$ (for more details see for example [72]). The physical interpretation of this solution is that the field free states are *dressed* by the photons of the laser field. In an experiment involving intensities around 10^{14} W cm^{-2}, the number of absorbed photons is very high such that $n = N \pm m$ with $m = 0, 1, 2, \ldots$ and N on the order of 10^{10} [72]. The resulting sets of parallel photon-dressed PECs may cross at a certain internuclear distance, as is shown in Fig. 2.12. Here, the two states have exactly the same energy due to the photon dressing. At this internuclear distance the states are thus resonant by the difference of the dressed number of photons. When this coupling between the states is neglected, the curves cross and are called *diabatic*. The diabatic curves have the shape of the field-free unperturbed curves. However at a resonant crossing, the radiative coupling via the laser field mixes the states—an effect that increases with the laser intensity and decreases with the order of the coupling [72]. The mixed states must appear as off-diagonal terms in the total Hamiltonian. For obtaining the perturbed PECs, the Hamiltonian thus needs to be diagonalised. These coupled PECs are called *adiabatic* curves and do not cross anymore. Instead, a gap opens with a width that increases with the coupling strength [72].

Even though based on a number of approximations, the adiabitic PECs of field-dressed states offer a rather intuitive picture for understanding a number of dissociation mechanisms, as will be shown briefly in the following Sect. 2.2.4.

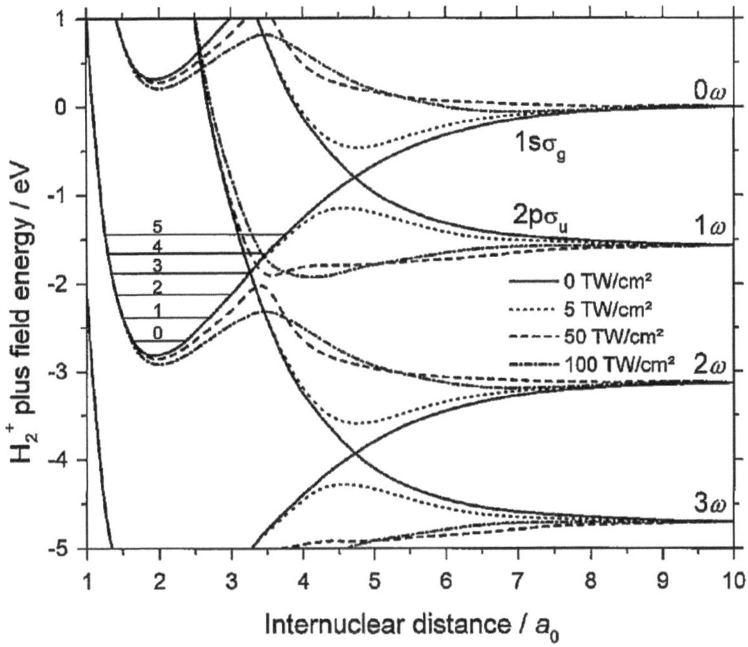

Fig. 2.12 PECs of H_2^+ dressed by a 790 nm laser field. When the field is off, the diabatic representation is appropriate and the *curves* cross (*solid line*). With increasing field strength, avoided crossings are formed at the resonance positions and the resonance gaps become visible in the adiabatic representation (Reprinted with permission from [73]. Copyright (1999) by the American Physical Society)

2.2.4 Dissociation Mechanisms

There are several purely laser induced pathways that may lead to the dissociation of a molecule. On the one hand, the laser field may induce transitions to dissociative states or vibrational states above the dissociation limit. On the other hand, the PECs can be dressed and deformed in such a way that dissociation is either assisted, reduced or delayed in time.

Field-free system

Part (a) of Fig. 2.13 shows the three possible pathways for light induced dissociation in the perturbative regime. (1) shows an electronic transition to a repulsive state. Due to the Coulombic repulsion, these curves are very steep for small internuclear distances, especially in the case of molecular ions with a high charge state. The dissociation products or fragments can thus acquire large kinetic energies E_{kin} as is illustrated in the figure. This dissociation mechanism is called *Coulomb explosion* and is described in more detail below. (2) depicts an electronic transition to an unbound vibrational state, as it lies above the dissociation limit. (3) illustrates a

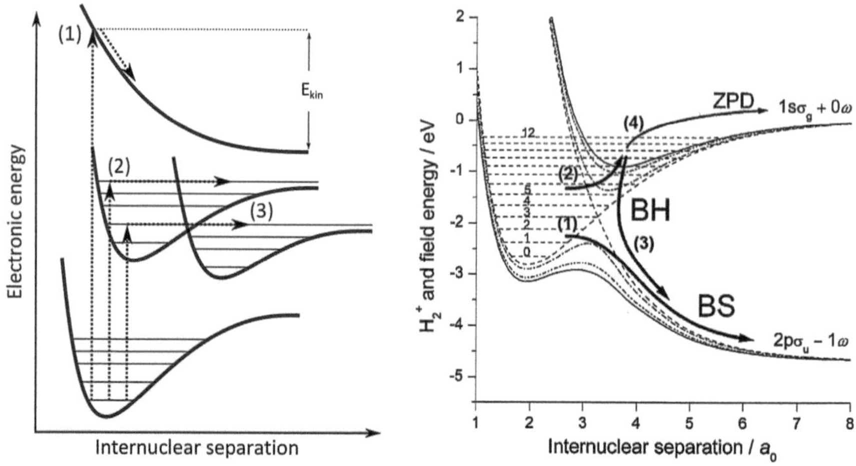

Fig. 2.13 **a** Dissociation pathways in a field free system: transition to a repulsive state (*1*), transition to an unbound state above the dissociation limit (*2*) and predissociation (*3*). **b** Dissociation pathways in a field-dressed system. The figure illustrates the possible dynamics at the 1-photon resonance in H_2^+, dressed by 266 nm photons and a maximum intensity of 5×10^{13} W cm^{-2} (*solid line*, largest gap). Details can be found in text (Part **b** reprinted with permission from [74]. Copyright (2001) by the American Physical Society)

process called *predissociation*. Here, a transition to a bound excited state takes place initially. However, if this excited state is coupled to a dissociative state on another PEC, the molecule may still fragment. Such couplings can internally be established via spin-orbit effects and conical intersections between PECs or externally by collisions for example [66]. Note that in pathways (2) and (3), the excess energy above the relevant dissociation limit is usually on the order of the vibrational level spacing and thus very small compared to (1). The corresponding fragments can typically assumed to be released with a thermal energy distribution centred at zero kinetic energy.

Field-dressed system

Dissociation mechanisms in a field-dressed system can be understood by following the movement of vibrational state distributions along the diabatic and adiabatic PECs. This has been studied intensively in the case of H_2^+ (for reviews see [69, 72]). The most prominent processes are *bond-softening* (BS), *bond-hardening* (BH) and *zero-photon dissociation* (ZPD). They are illustrated in part (b) of Fig. 2.13 in the case of the 1-photon crossing of the bonding $1s\sigma_g$ and the antibonding $2p\sigma_u - 1\omega$ state. At the leading edge of the laser pulse, the intensity is very low and the radiative coupling can be neglected. Here, the diabatic PECs apply (dashed line). When the laser intensity is increased, the 1-photon gap opens at the resonant internuclear distance (the highest intensity corresponds to the solid line). The gap reduces the bound state to a shallow potential barrier. Population in vibrational states higher than this barrier can thus escape via pathway (1) which corresponds to BS. This process was first observed experimentally by Bucksbaum et al. [75].

At the same time however, the anti-bonding PEC turns into a bound state, due to the potential minimum formed by the 1-photon gap. This means that population in vibrational states above this minimum can be trapped, which is denoted by pathway (2). From here, two different dissociation pathways are possible. Pathway (3) denotes the first, BH, which was first predicted by Giusti-Zusor and Mies [76] and confirmed experimentally only 7 years later by Frasinski et al. [73]. Here, the trapped vibrational population dissociates at the falling edge of the laser pulse, when the gap reduces and the potential minimum turns into a repulsive curve again. Note that the energy of the system is reduced by one photon during this transition. This excess energy is shared by the two fragments in form of kinetic energy. Pathway (4) shows the second dissociation pathway for the trapped vibrational population via ZPD [74]. If the laser intensity increases even further than shown in the figure, the gap becomes larger and flattens out the upper potential minimum and the trapped population becomes unbound. As the population remains on the $1s\sigma_g$ curve throughout this dissociation pathway no photon absorption or emission is involved, hence the name ZPD.

Coulomb explosion

If the electrons of the molecule are stripped off instantaneously with respect to the period of nuclear motion, the internuclear geometry is preserved during this process. The resulting Coulomb repulsion between the remaining nuclii is then given by:

$$\mathbf{F} = \frac{1}{4\pi\epsilon_0} \sum_i - \frac{ZZ_i e^2}{|\mathbf{R} - \mathbf{R}_i|^3} (\mathbf{R} - \mathbf{R}_i) \tag{2.39}$$

Therefore, if one measures the momentum vector of each ejected fragment after the Coulomb explosion, one may use the above equation to reconstruct their initial positions and hence the internuclear geometry, alignment and orientation of the molecule at the instant of explosion [77]. This can be achieved by employing laser pulses with durations significantly shorter than the vibrational period of the molecule in question and intensities generally $>10^{14}$ W cm^{-2} [77, 78]. In the PEC picture, this process corresponds to a vertical transition to a repulsive- or Coulombic-curve.

In this process, the pulse duration is the main critical factor for preserving the internuclear structure during the ionisation process. I_2 for example has a vibrational period of roughly 150 fs and hence a 20 fs laser pulse can be used for Coulomb explosion [77]. Furthermore high laser intensities should be employed in order to reach very high charge states. This increases the Coulombic repulsion between the fragments and thus accelerates the explosion process. However, non-adiabatic or post-ionisation alignment may take place in a highly intense laser pulse, which will be discussed in Sect. 2.4.2. While these effects may partially be minimised in the analysis of the fragment momenta, they may be fully suppressed by using circularly polarised light. In this case there is no preferred polarisation direction and the molecules remain randomly aligned in the polarisation plane.

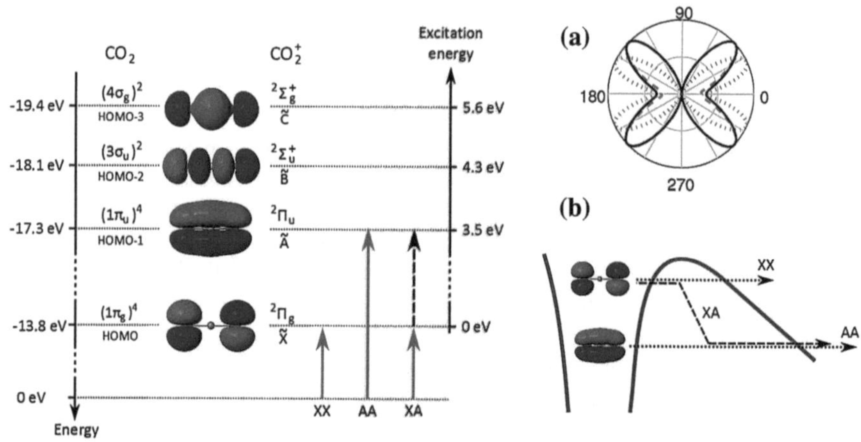

Fig. 2.14 The four highest occupied orbitals in CO_2 and the associated excited ionic states. Three ionisation channels are shown: direct tunneling from HOMO resulting in the ionic ground state (*XX*), tunneling from HOMO-1 (*AA*) and inelastic tunneling from HOMO correlating with HOMO-1. Inset **a** shows the MO-ADK rate from the HOMO (*dotted line*) compared to ab initio solutions of the TDSE using the single active electron approximation and 800 nm laser pulses. The *dashed* and *solid lines* correspond to intensities of 5.6×10^{13} W cm^{-2} and 1.1×10^{14} W cm^{-2}, respectively. **b** Illustrates the three ionisation channels in the inelastic tunneling picture (Part **a** reprinted with permission from [79]. Copyright (2009) by the American Physical Society)

Dissociation in the strong field regime

In the strong field regime, transitions to dissociative states may also take place through tunnel ionisation and inelastic recollisions. This is discussed in more detail in Sects. 2.3.2 and 2.5.

2.3 Molecular Tunneling Ionisation

Due to their non-spherical electronic orbitals and nuclear degrees of freedom, molecular ionisation has additional features compared to atomic ionisation. The most important ones are reviewed in this section with a particular emphasis on CO_2.

2.3.1 *Orbital Structure Effects*

The rich nuclear structure of molecules leads to orbitals with different symmetry properties, as was discussed in Sect. 2.2.1. One may thus expect that the TI rate in the molecular frame is anisotropic. This is taken into account in the molecular Ammosov-Delone-Krainov (MO-ADK) theory [80] . In this framework, the essential features of the angular ionisation rate are obtained under the assumption that the ionisation

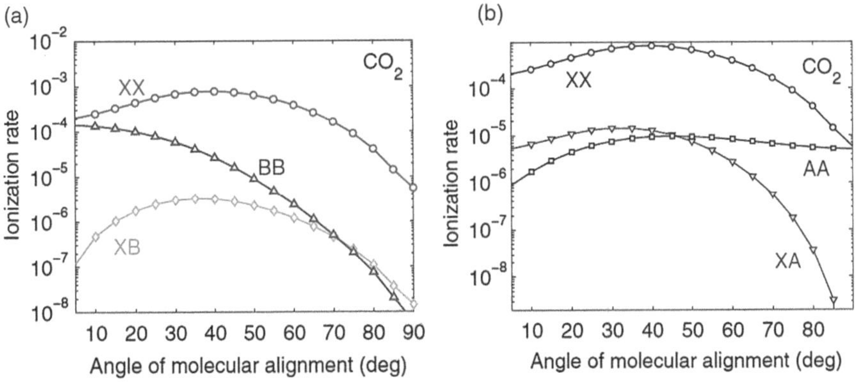

Fig. 2.15 Calculated relative angular ionisation rates for the available channels in CO_2 involving the highest three orbitals. The employed laser pulse parameters are $\lambda = 800\,\text{nm}$ and $I = 0.8 \times 10^{14}\,\text{W cm}^{-2}$. The calculations include the direct channels *XX*, *AA* and *BB* and the inelastic tunneling channels correlating with HOMO; *XA* and *XB* (Reprinted with permission from [89]. Copyright (2012) by the American Physical Society)

probability scales with the electronic density profile of the respective molecular orbital along the ionising laser polarisation direction. This scenario is complicated further by the fact that the occupied orbitals in a molecule typically have different symmetries (see Fig. 2.14). In an experiment, one would observe the superposition of the angular TI rates from all orbitals. However, due to the exponential decay of the TI rate with the ionisation potential I_p, ionisation from the highest occupied molecular orbital (HOMO) is expected to dominate. Yet, it has been observed that the influence of lower lying orbitals cannot be excluded a priori [81–84] - especially not in CO_2. Taking these additional TI channels into account is sometimes referred to as *multichannel strong field ionisation* [84]. Angular ionisation rates for CO_2 from its three highest orbitals obtained in such a framework are displayed in Fig. 2.15.

It should also be added, that experimental ionisation studies in aligned CO_2 show quantitative discrepancies with the MO-ADK model (see inset (a) in Fig. 2.14) [85], suggesting significant shortcomings (alternative approaches and modifications can be found in [79, 86–89] for example). For a qualitative understanding of the experimental results in this thesis however, it is a reasonable tool.

Typically, molecules have decreasing I_p with increasing size. This is associated with the intuitive picture of the outermost electron being further away from the positively charged nuclear framework and thus more loosely bound. However, it was found that most organic molecules have a higher saturation intensity I_{sat} than predicted by their TI rate [90, 91]. Several sources may contribute to this apparent resistance to strong field ionisation. The reduction of the TI rate due to nodal planes in the molecular orbital has already been introduced and is probably the most prominent cause.

2.3.2 *Tunneling Excitation*

Through TI, the molecular ion may end up in an excited state. Here, two pathways are possible: TI from lower lying orbitals and inelastic tunneling. Both channels may thus lead to the dissociation of the molecular ion.

TI from lower lying orbitals

Previously, TI was presented as the removal of an electron from a given orbital. In the single active electron approximation, the evolution of the remaining electrons is neglected. This is implicit in Koopmans' theorem that says that the remaining electrons do not adjust their distribution during ionisation [92]. This implies that excited states in the parent ion can be reached through ionisation from lower lying orbitals (see Fig. 2.14). Furthermore the ionic state has the same orbital symmetry as its associated state in the neutral. In the molecular frame, the probability of reaching an excited state via this channel thus scales like the MO-ADK rate for the associated neutral orbital. The relative strength of this channel can thus be controlled by the laser polarisation with respect to the molecular axis.

Inelastic tunneling

The breakdown of the SAE has already been mentioned in Sect. 2.1.2. When allowing for nonadiabatic electron dynamics during TI, the different channels may interact. This leads to excitation and deexcitation of the ion and is associated with the electron hole moving to another orbital during TI [89]. This mechanism is thus sometimes referred to as *inelastic tunneling*. It is illustrated for the XA channel in inset (b) in Fig. 2.14. Here, TI from HOMO takes place initially. During tunneling, the hole moves to HOMO-1 leaving the ion in the first excited state A. In this way, the correlation channel does not possess the full exponential suppression of the direct channel (TI from HOMO-1, AA). It has therefore been suggested that inelastic tunneling becomes most relevant for low lying orbitals and low intensity laser fields. Torlina et al. [89] have recently calculated the angular dependence of such correlated channels in CO_2. The results are displayed in Fig. 2.15. It should be pointed out, however, that experimental evidence concerning inelastic tunneling is currently not available.

2.4 NSDI in Small Molecules

As in atomic NSDI, the recollision model also applies to NSDI in small molecules. Similar to the above descriptions, characteristic ion yield enhancements ('knee structures') [93], polarisation dependence [93, 94] and electron-electron correlations in ion momentum spectra from COLTRIMS studies [39, 51, 95] have also been reported in small molecules like H_2, D_2, N_2, O_2 and CO_2. Therefore the overall role of the laser parameters discussed in the Sect. mol:ionmol:exc also holds true for small molecules.

The work by Guo and coworkers [93, 96, 97] serves as a useful reference for the intensity regime in which NSDI in N_2, O_2 and CO_2 can be observed and the

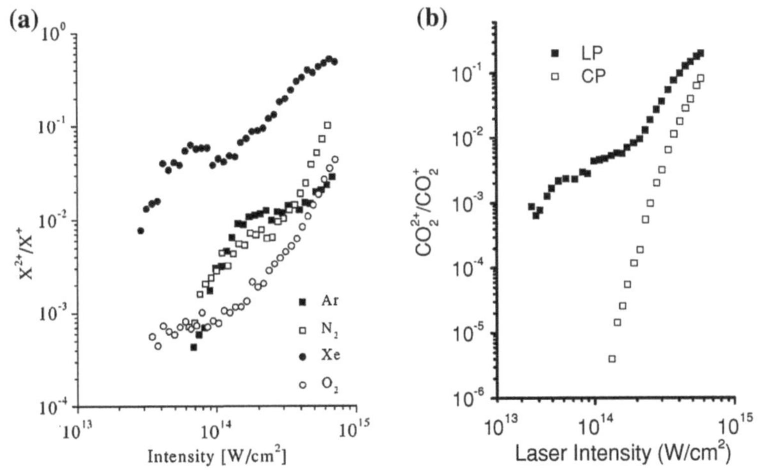

Fig. 2.16 Part **a** displays the X^{2+}/X^{+} ratio explained in the text for N_2 and O_2 and their reference atoms with similar I_p Ar and Xe. The region with weak intensity dependence identifies NSDI suggesting an intensity regime of roughly $1.5-4\times10^{14}\,\mathrm{W\,cm^{-2}}$ for N_2 and $0.4-1\times10^{14}\,\mathrm{W\,cm^{-2}}$ for O_2 at 800 nm. Part **b** shows the same plot for CO_2 using linear (*LP*) and circular polarisation (*CP*) indicating an NSDI regime for $0.5-2\times10^{14}\,\mathrm{W\,cm^{-2}}$ at 800 nm (Reprinted with permission from [93] and [97]. Copyright (1998) and (2010) by the American Physical Society)

corresponding plots are displayed in Fig. 2.16. In this case, NSDI was identified by the intensity dependence of the ratio X^{2+}/X^{+}, where X refers to the molecule. For a nonsequential process X^{+} is an intermediate state. Therefore the intensity dependence of the above ratio should be weak. For sequential ionisation on the other hand, X^{+} is a precursor state and hence the corresponding ratio should have a strong intensity dependence as this would lead to a depletion of the X^{+} yield along with and increase of the X^{2+} yield. Furthermore, the results show that in N_2 NSDI is pronounced the most and in O_2 the least. The authors attribute this to the open shell structure of O_2, with the two outermost electrons occupying different but degenerate orbitals. This would lead to a lower probability of removing both in a non-sequential process. CO_2 on the other hand has a closed shell structure (the outermost $1\pi_g$ orbital is fully filled with four electrons), but the two electrons can either be removed from different (degenerate) or the same orbital [97]. Following the author's arguments, this would explain the higher NSDI rate compared to O_2 and lower compared to N_2, where the two electrons will always come from the same fully filled ($3\sigma_g^2$) orbital.

Due to their rich nuclear structure, several additional laser-molecule interactions can take place and restrain the observation of the different NSDI channels. The most important ones in the context of this thesis are the influence of the molecular orbital asymmetry on the first ionisation step and dynamic alignment effects. Available experimental results are briefly reviewed in the following sections.

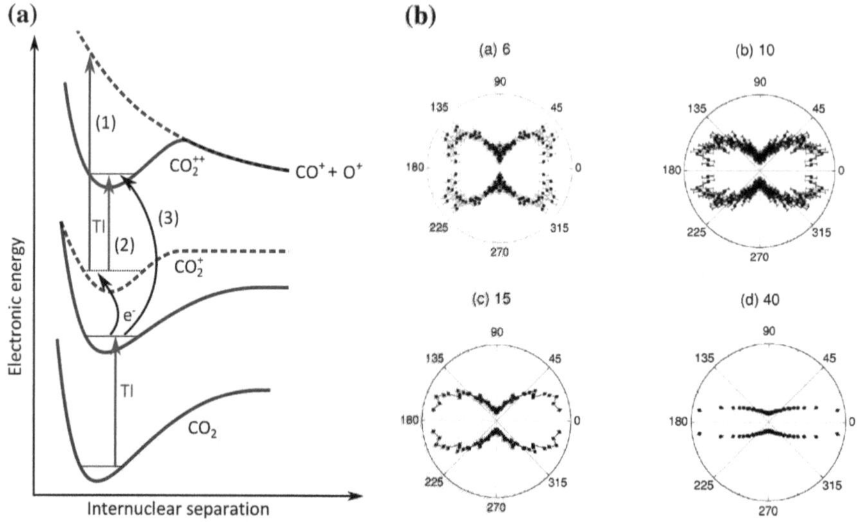

Fig. 2.17 **a** Graphical representation of possible pathways in molecular NSDI with CO_2 as example. *Dashed* PECs denote excited states, TI refers to tunneling ionisation and e^- to recollision. (*1*) RESI to a repulsive state of CO_2^{++} resulting in two charged fragments. (*2*) RESI to a bound state of CO_2^{++}. (*3*) (e, 2e) channel to a bound CO_2^{++} state. **b** Angularly resolved relative yields of channel (*1*) from the *left panel*. For all measurements the laser polarisation is *horizontal* and for each plot the estimated peak intensity is denoted in 10^{13} W cm^{-2} (Part **b** reprinted with permission from [99]. Copyright (2005) by the American Physical Society)

2.4.1 Molecular Structure Effects

The crucial effect of the molecular geometry and associated orbitals is its impact on ionisation probability, as mentioned already in Sect. 2.3.1. In particular, Alnaser et al. [98, 99] measured the angular distributions of ion fragments, created in a dissociative double ionisation of several di- and triatomic molecules. Here, CO_2 is treated as an example. The authors focussed on dissociation channels with two singly charged fragments in order to reconstruct the initial molecular alignment from the fragment's momentum vectors detected in coincidence. This is illustrated as channel (1) in part (a) of Fig. 2.17, which also displays the RESI and (e, 2e) channels in the molecular potential energy curve picture. Part (b) of Fig. 2.17 then shows the experimental angular distributions for different intensities. Here, 8 fs pulses at 790 nm were employed in the NSDI intensity regime. At the lowest intensity a clear maximum at approximately 40° with respect to the laser polarisation along the y-axis can be observed. This agrees well with the angular distribution of TI in CO_2 (see Fig. 2.14) and thus suggests that it is mainly the angular dependence from the first ionisation step that determines the angular distribution of the overall process. Similar results were found for the other molecules and are in very good agreement with theoretical predictions and earlier studies [95].

2.4.2 Laser Induced Alignment Effects

There are two types of alignment of the molecular geometry in the laboratory frame, which can alter the angular distribution of ion fragments: *dynamic alignment* (DA) or non-adiabatic alignment (for details see Chap. 5) describing the molecule's laser induced reorientation before the dissociation, due to its dipole moment and *post-ionisation alignment* (PIA), which describes a deflection of the ion fragments in the presence of the laser pulse due the molecular polarizability [100, 101].

Both effects depend on an interplay between the polarisability or induced dipole moment of the molecule, its mass and the pulse length (at a given intensity). For heavy molecules like I_2 DA can generally be neglected [100, 102]. For lighter molecules, Alnaser et al. [99] showed that the effect can be neglected for example in CO_2 for short pulses around 8 fs and low intensities at 6×10^{13} W cm^{-2}. This was confirmed by a study by Tong [101] and Voss [103] in O_2, which additionally showed that PIA cannot be neglected for 35 fs pulses but does not play a role at 8 fs and low intensities. While it is not easy to disentangle the DA and PIA, their impact can be observed in the angular fragment distributions in part (b) of Fig. 2.17. For increasing intensity, the distribution becomes increasingly confined around the laser polarisation direction.

2.4.3 NSDI in Aligned Diatomics

In order to determine the influence of the molecular structure on the recollision step, one has to disentangle it from the ionisation and propagation step, which is possible when performing measurements on NSDI in the molecular frame. Using COLTRIMS, the molecular frame is reconstructed from the momentum vectors of the fragments. It is thus not possible to employ this approach when detecting doubly ionised molecules directly. For this, Zeidler et al. [39] used COLTRIMS together with non-adiabatic alignment to probe the molecules in their frame of reference. It was found that for the molecule being aligned along the laser polarisation, NSDI is slightly more probable and the RESI channel less pronounced. One important aspect of Zeidler's experiment is a possible route to the removal of the ionisation step by determining the single ionisation yield of N_2 prior to the main experiment and thus determining the NSDI yields as ratios of double and single ionisation yields. In order to take any kind of fluctuations into account, the relative alignment-dependent single ionisation yields were determined from a Ne/N_2 gas mixture, where the Ne yield works as a reference. In this way 19 % more N_2^+ and 30 % more N_2^{++} ions were observed for parallel molecules at $(1.2 \pm 0.2) \times 10^{14}$ Wcm^{-2}. This is in good agreement with a similar experiment performed by Litvinyuk et al. where the N_2^+ yield was 23 % higher for parallel aligned molecules at 2×10^{14} Wcm^{-2} [104].

Zeidler's findings are in agreement with several theoretical studies [83, 105, 106]. Jia et al. [83] use an S-Matrix approach to calculate alignment-dependent ratios of N_2^{++}/N_2^+ taking into account the initial ionisation probability and the probability of the first ionised electron being driven back to the ion. The authors attribute the alignment-dependence of the NSDI yield to the corresponding dependence of the ionisation cross section of N_2^+. The authors suggest that the angularly resolved recollision cross section scales like the electron density profile of the parent ion. Furthermore, the authors also find that the influence of lower lying orbitals must be included to obtain a better agreement with the experimental data.

2.5 Recollision Induced Fragmentation

Recollision induced fragmentation (RIF) takes place when the molecular ion is excited to a repulsive state via electron impact ionisation or excitation. RIF can be divided into a dissociative excitation (DE) and dissociation ionisation channel (DI) (see channel (1) in part (a) of Fig. 2.17). In DE the parent ion is excited to a dissociative state (see part (a) in Fig. 2.18), whereas in DI the parent ion is ionised further, such that the higher charged ion fragments. The recollision contribution to such mechanisms is studied significantly less intensively than NSDI, but its qualitative behaviour under a change of laser parameters is expected to be similar. In this section, experimental results obtained in small and large molecules are briefly reviewed.

2.5.1 *Small Molecules*

DE has so far been observed in CO_2 [27] and N_2 [26] and O_2 [107]. DI has been observed in N_2 and O_2 [96, 107, 108]. In the case of CO_2 McKenna et al. suggest that RIF is responsible for a significant DE enhancement for linear compared to circular polarisation [27]. The results are displayed in part (b) of Fig. 2.18. Here, recollision is switched off for circular polarisation, while the intensity is adjusted to match the effective ionisation rate for both polarisation states (see Sect. 2.1.2). With the same arguments, RIF could be identified in the (1, 0) channel in N_2 [26], even though the results are probably only reliable at low intensities, where molecular alignment effects of the driving laser field were weak.

The angular distributions of DE and DI in O_2 were studied recently with a COLTRIMS [107]. As expected, the distributions qualitatively follow the single ionisation rate distribution, yet DE is peaked around 30° and DI around 40°. This is an interesting result pointing towards a significant influence of the recollision step in these processes.

Wu et al. [108, 109] have studied the polarisation dependence of the DI channel in N_2 and O_2 in passively aligned molecules. This means that in the experiment the detection geometry restricted the collection to molecules aligned along one axis in the laboratory frame. Recollision was identified via the intensity dependence of the X^{2+}/X^{+} ratio explained in the Sect. 2.5 and compared for parallel and perpendicular polarisation with respect to the molecular bond. It was found that for parallel polarisation, recollision induced DI is significantly suppressed in O_2.

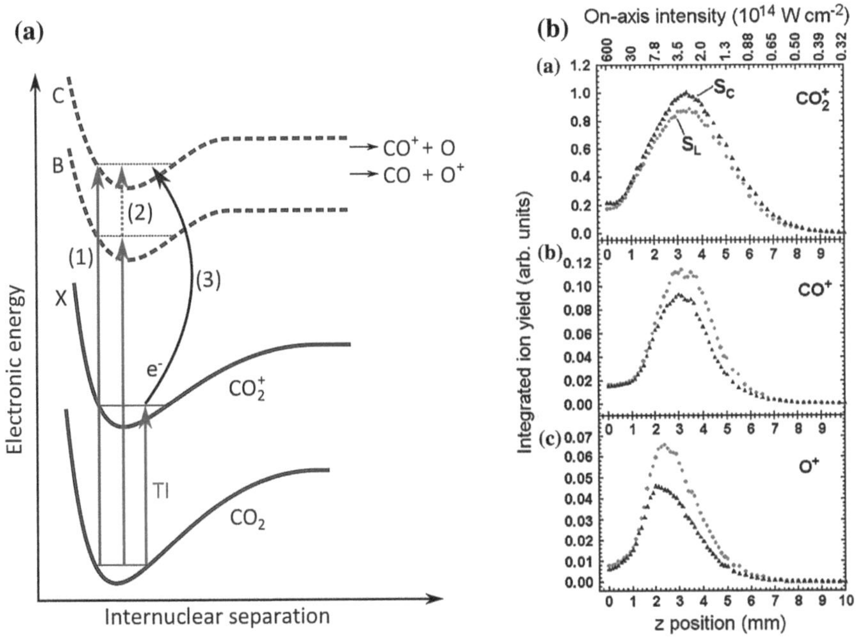

Fig. 2.18 **a** possible dissociative excitation pathways in CO_2. *Dashed* PECs denote excited states, TI refers to tunneling ionisation and e^- to recollision. Note that only the C state is above the relevant dissociation limits. (*1*) direct TI, (*2*) TI to B state and photoexciation (*dotted arrow*) to C state, (*3*) TI followed by recollision induced excitation to C state. **b** Ion yields in CO_2 as a function of intensity for linear (*red circles*) and circular (*blue triangles*) polarisations using 55 fs, 790 nm pulses. Even though the single ionisation rate is not perfectly matched, a clear enhancement for linear polarisation is visible (Part **b** reprinted with permission from [27]. Copyright (2006) by the American Physical Society)

2.5.2 Large Molecules

RIF has been observed as a significant mechanism in the alkanes methane, ethane and propane [110–112], water [112], in the relatively small organic molecule benzene [112, 113], in a series of alcohols [114] and in C_60 [115]. Yet its contribution has also been shown to be negligible in typical organic molecules like anthracene [116] and naphtalene [117].

In the experiment by Rajgara et al. [114] a series of alcohols of increasing size was investigated using pulses at 800 nm, 100 fs, 10^{16} W cm^{-2}. Recollision was shown to be the dominant fragmentation mechanism by observing a significant suppression of the fragmentation yields, when using circular polarisation. The high intensity was chosen, as experiments on benzene by Talebpour et al. with pulses at 800 nm, 100 fs, 10^{15} W cm^{-2} did not show such a suppression and hence ruled out recollision as a mechanism [118]. Other than investigating this threshold intensity (and thus recollision energy) the impact of the molecular structure of the differently sized

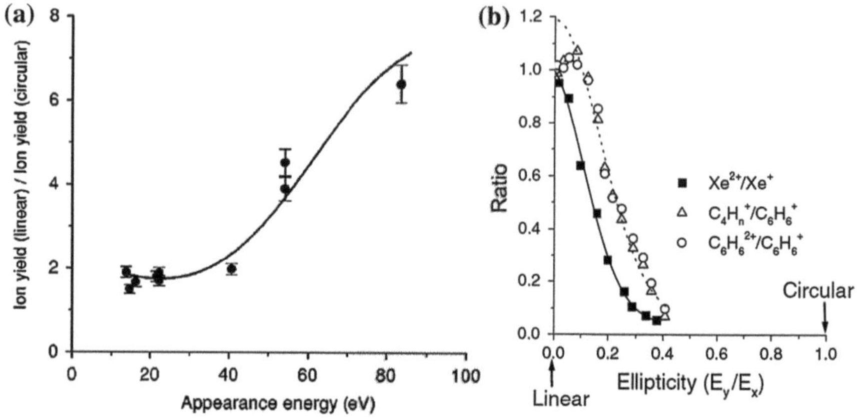

Fig. 2.19 **a** Impact of the molecular structure in differently sized alkohols on recollision induced fragmentation. Here, the ratio of fragment ion yield at *linear* to the yield at *circular* polarisation is displayed as a function of appearance energy. This energy is a measure of the energy transfer to the molecule required to produce the corresponding fragment ion. **b** Ellipticity dependence of fragmentation of benzene (Part **a** reproduced with permission from [114]. Copyright (2003), AIP Publishing LLC. Part **b** reprinted with permission from [113]. Copyright (2001) by the American Physical Society)

alcohols was investigated. By measuring the relative suppression of several fragments corresponding to different appearance energies, the authors concluded that suppression and hence recollision induced fragmentation was most dominant for the fragmentation channels which require the highest energy transfer to the cation. The corresponding data is displayed in part (a) of Fig. 2.19.

In another series of measurements with the same setup, Rajgara and coworkers confirmed these results in the molecules water, methanol and benzene [112]. In addition to the above measurements at very high intensity, a lower bound of $2.0 \times 10^{15}\,\mathrm{W\,cm^{-2}}$ (379 eV) was found for recollision induced fragmentation in benzene. This is only marginally higher than in Talebpour's experiment and explains, why it was not observed then. It should be noted however that a few details remain open in the above experiments. It is for example not clear how big the impact of the peak intensity variation on the recorded yields is, when switching from linear to circular polarisation. This is because the intensity was kept constant throughout the experiment. The authors also point out to be above the saturation intensity of the investigated molecules, which means that the effective peak intensity acting on the recolliding electron remains unknown. Furthermore the fragmentation event upon recollision was not investigated further, such that it is not clear whether an excitation to a dissociative state or an impact fragmentation was present.

Bhardwaj et al. [113] then observed recollision induced fragmentation in benzene by increasing the recollision energy to 401 eV via a longer wavelength of 1400 nm, compared to 189 eV in Talebpour's experiment [118]. This has the advantage of staying below the saturation intensity of the molecule. The pulse length was 70 fs, with an intensity of $7 \times 10^{14}\,\mathrm{W\,cm^{-2}}$ on target. Recollision induced fragmentation was identified by observing the same ellipticity dependence for the fragmentation and NSDI channels. The corresponding data is displayed in part (b) of Fig. 2.19. Furthermore the fragmentation and NSDI yields have an unexpected maximum at an ellipticity of about 0.1. This is interpreted by the authors as follows. A nodal plane aligned along the linear laser polarisation direction leads to a recolliding wavepacket with non-zero transverse momentum, driving it away from the polarization direction and hence the parent ion. However, in an elliptical driving laser field, the transverse field component may compensate for the electron's initial momentum. This manifests itself in a maximum NSDI yield for nonzero laser ellipticity. Due to the large number of broad nodal planes in the HOMO of C_6H_6, this effect is expected to be visible even in a randomly oriented sample. A similar explanation has been used to explain the absence of recollision induced effects in CS_2 [119].

References

1. A. Einstein, Über einen die erzeugung und verwandlung des lichtes betreffenden heuristischen gesichtspunkt. Annalen der Physik **322**(6), 132–148 (1905)
2. G.S. Voronov, N.B. Delone, Sov. Phys. JETP Lett. **1**, 66 (1965)
3. C.H.K.M Protopapas, P.L. Knight, Atomic physics with super-high intensity lasers. Rep. Prog. Phys. **60**, 389 (1997)
4. P. Agostini, F. Fabre, G. Mainfray, G. Petite, N.K. Rahman, Free-free transitions following six-photon ionization of xenon atoms. Phys. Rev. Lett. **42**, 1127–1130 (1979)
5. F. Fabre, G. Petite, P. Agostini, M. Clement, Multiphoton above-threshold ionisation of xenon at 0.53 and 1.06 μ m. J. Phys. B: At. Mol. Phys. **15**(9), 1353–1369 (1982)
6. T. Brabec, *Strong Field Laser Physics* (Springer, Berlin, 2008)
7. L.A. Lompré, A. L'Huillier, G. Mainfray, C. Manus, Laser-intensity effects in the energy distributions of electrons produced in multiphoton ionization of rare gases. J. Opt. Soc. Am. B: Opt. Phys. **2**, 1906–1912 (1985)
8. G.G. Paulus, W. Nicklich, H. Walther, Investigation of above-threshold ionization with femtosecond pulses: connection between plateau and angular distribution of the photoelectrons. Europhys. Lett. (EPL) **27**, 267–272 (1994)
9. G.G. Paulus, W. Nicklich, H. Xu, P. Lambropoulos, H. Walther, Plateau in above threshold ionization spectra. Phys. Rev. Lett. **72**, 2851–2854 (1994)
10. L. Keldysh, Ionization in the field of a strong electromagnetic wave. Sov. Phys. JETP **20**, 1314, 1307 (1965)
11. T. Brabec, F. Krausz, Intense few-cycle laser fields: frontiers of nonlinear optics. Rev. Mod. Phys. **72**, 545 (2000)
12. F. Krausz, M. Ivanov, Attosecond physics. Rev. Mod. Phys. **81**, 163 (2009)
13. G.L. Yudin, M.Y. Ivanov, Nonadiabatic tunnel ionization: looking inside a laser cycle. Phys. Rev. A **64**, 013409 (2001)
14. F.H.M. Faisal, Multiple absorption of laser photons by atoms. J. Phys. B: At. Mol. Phys. **6**, L89–L92 (1973)
15. H.R. Reiss, Effect of an intense electromagnetic field on a weakly bound system. Phys. Rev. A **22**, 1786–1813 (1980)
16. M. Lewenstein, P. Balcou, M.Y. Ivanov, A. L'Huillier, P.B. Corkum, Theory of high-harmonic generation by low-frequency laser fields. Phys. Rev. A **49**, 2117 (1994)
17. A.M. Perelomov, V.S. Popov, M.V. Terent'ev, Ionization of atoms in an alternating electric field. Sov. J. Exp. Theor. Phys. **23**, 924 (1966)
18. A.M. Perelomov, V.S. Popov, M.V. Terent'ev, Ionization of atoms in an alternating electric field: II. Sov. J. Exp. Theor. Phys. **24**, 207 (1967)

19. A.M. Perelomov, V.S. Popov, Ionization of atoms in an alternating electrical field. III. Sov. J. Exp. Theor. Phys. **25**, 336 (1967)
20. N.B.D.M.V. Ammosov, V.P. Krainov, Tunnel ionization of complex atoms and atomic ions by an alternating electromagnetic field. Sov. Phys. JETP **64**, 1191–1194 (1986)
21. S.F.J. Larochelle, A. Talebpour, S.L. Chin, Coulomb effect in multiphoton ionization of rare-gas atoms. J. Phys. B: At. Mol. Opt. Phys. **31**, 1215–1224 (1998)
22. P.B. Corkum, Plasma perspective on strong field multiphoton ionization. Phys. Rev. Lett. **71**, 1994–1997 (1993)
23. C. Cornaggia, P. Hering, Nonsequential double ionization of small molecules induced by a femtosecond laser field. Phys. Rev. A **62**(2), 023403 (2000)
24. M.Y. Ivanov, M. Spanner, O. Smirnova, Anatomy of strong field ionization. J. Mod. Opt. **52**, 165–184 (2005)
25. M. Suresh, J. McKenna, B. Srigengan, I. Williams, E. English, S. Stebbings, W. Bryan, W. Newell, E. Divall, C. Hooker, A. Langley, Multiple ionization of ions and atoms by intense ultrafast laser pulses. Nucl. Instrum. Methods Phys. Res. Sect. B **235**(1–4), 216–220 (2005)
26. J. McKenna, M. Suresh, B. Srigengan, I.D. Williams, W.A. Bryan, E.M.L. English, S.L. Stebbings, W.R. Newell, I.C.E. Turcu, J.M. Smith, E.J. Divall, C.J. Hooker, A.J. Langley, J.L. Collier, Ultrafast ionization study of N_2 in intense linearly and circularly polarized laser fields. Phys. Rev. A **73**(4), 043401 (2006)
27. J. McKenna, M. Suresh, B. Srigengan, I.D. Williams, W.A. Bryan, E.M.L. English, S.L. Stebbings, W.R. Newell, I.C.E. Turcu, J.M. Smith, E.J. Divall, C.J. Hooker, A.J. Langley, J.L. Collier, Rescattering-enhanced dissociation of a molecular ion. Phys. Rev. A **74**, 043409 (2006)
28. B. Zon, Tunneling ionization of atoms with excitation of the core. J. Exp. Theor. Phys. **91**(5), 899–904 (2000)
29. M. Lezius, V. Blanchet, D.M. Rayner, D.M. Villeneuve, A. Stolow, M.Y. Ivanov, Nonadiabatic multielectron dynamics in strong field molecular ionization. Phys. Rev. Lett. **86**, 51 (2001)
30. W.A. Bryan, S.L. Stebbings, J. McKenna, E.M.L. English, M. Suresh, J. Wood, B. Srigengan, I.C.E. Turcu, J.M. Smith, E.J. Divall, C.J. Hooker, A.J. Langley, J.L. Collier, I.D. Williams, W.R. Newell, Atomic excitation during recollision-free ultrafast multi-electron tunnel ionization. Nat. Phys. **2**(6), 379–383 (2006)
31. A.N. Pfeiffer, C. Cirelli, M. Smolarski, R. Dörner, U. Keller, Timing the release in sequential double ionization. Nat. Phys. **7**(5), 428–433 (2011)
32. A.N. Pfeiffer, C. Cirelli, M. Smolarski, X. Wang, J.H. Eberly, R. Drner, U. Keller, Breakdown of the independent electron approximation in sequential double ionization. New J. Phys. **13**, 093008 (2011)
33. I.V. Litvinyuk, F. Logar, P.W. Dooley, D.M. Villeneuve, P.B. Corkum, J. Zanghellini, A. Pegarkov, C. Fabian, T. Brabec, Shakeup excitation during optical tunnel ionization. Phys. Rev. Lett. **94**, 033003 (2005)
34. Z.B. Walters, O. Smirnova, Attosecond correlation dynamics during electron tunnelling from molecules. J. Phys. B: At. Mol. Opt. Phys. **43**(16), 161002 (2010)
35. G.G. Paulus, W. Becker, W. Nicklich, H. Walther, Rescattering effects in above-threshold ionization: a classical model. J. Phys. B: At. Mol. Opt. Phys. **27**, L703–L708 (1994)
36. M. Lein, Molecular imaging using recolliding electrons. J. Phys. B: At. Mol. Opt. Phys. **40**(16), R135–R173 (2007)
37. K.C. Kulander, K.J. Schafer, J.L. Krause, Super-intense laser-atom physics, in *NATO Advanced Study Institute Series B: Physics*, vol. 316 (Plenum, New York, 1993), p. 95
38. R. Thomson, C. Leburn, D. Reid (eds.), *Ultrafast Nonlinear Optics* (Springer, Berlin, 2013)
39. D. Zeidler, A. Staudte, A.B. Bardon, D.M. Villeneuve, R. Dorner, P.B. Corkum, Controlling attosecond double ionization dynamics via molecular alignment. Phys. Rev. Lett. **95**, 203003 (2005)
40. K.S. Budil, P. Salires, M.D. Perry, A. L'Huillier, Influence of ellipticity on harmonic generation. Phys. Rev. A **48**, R3437 (1993)

41. J. Guo, T. Wang, X.-S. Liu, J.-Z. Sun, Non-sequential double ionization of mg atoms in elliptically polarized laser fields. Laser Phys. **23**, 055303 (2013)
42. T. Brabec, M.Y. Ivanov, P.B. Corkum, Coulomb focusing in intense field atomic processes. Phys. Rev. A **54**(4), R2551–R2554 (1996)
43. X. Wang, J.H. Eberly, Elliptical trajectories in nonsequential double ionization. New J. Phys. **12**, 093047 (2010)
44. J. Tate, T. Auguste, H.G. Muller, P. SaliTres, P. Agostini, L.F. DiMauro, Scaling of wave-packet dynamics in an intense midinfrared field. Phys. Rev. Lett. **98**, 013901 (2007)
45. S. Augst, D. Strickland, D.D. Meyerhofer, S.L. Chin, J.H. Eberly, Tunneling ionization of noble gases in a high-intensity laser field. Phys. Rev. Lett. **63**, 2212 (1989)
46. T. Auguste, P. Monot, L.A. Lompre, G. Mainfray, C. Manus, Multiply charged ions produced in noble gases by a 1 ps laser pulse at lambda = 1053 nm. J. Phys. B: At. Mol. Opt. Phys. **25**(20), 4181 (1992)
47. D.N. Fittinghoff, P.R. Bolton, B. Chang, K.C. Kulander, Observation of nonsequential double ionization of helium with optical tunneling. Phys. Rev. Lett. **69**, 2642 (1992)
48. U. Eichmann, M. Dörr, H. Maeda, W. Becker, W. Sandner, Collective multielectron tunneling ionization in strong fields. Phys. Rev. Lett. **84**, 3550 (2000)
49. E. Eremina, Electron correlation in multiple ionization of atoms and molecules by intense ultra-short laser pulses. Ph.D. thesis, TU Berlin (2005)
50. D.N. Fittinghoff, P.R. Bolton, B. Chang, K.C. Kulander, Polarization dependence of tunneling ionization of helium and neon by 120-fs pulses at 614 nm. Phys. Rev. A **49**, 2174 (1994)
51. O. Herrwerth, A. Rudenko, M. Kremer, V.L.B. de Jesus, B. Fischer, G. Gademann, K. Simeonidis, A. Achtelik, T. Ergler, B. Feuerstein, C.D. Schrter, R. Moshammer, J. Ullrich, Wavelength dependence of sub-laser-cycle few-electron dynamics in strong-field multiple ionization. New J. Phys. **10**(2), 025007 (2008)
52. B. Feuerstein, R. Moshammer, D. Fischer, A. Dorn, C.D. Schrter, J. Deipenwisch, J.R.C. Lopez-Urrutia, C. Hhr, P. Neumayer, J. Ullrich, H. Rottke, C. Trump, M. Wittmann, G. Korn, W. Sandner, Separation of recollision mechanisms in nonsequential strong field double ionization of Ar: the role of excitation tunneling. Phys. Rev. Lett. **87**, 043003 (2001)
53. R. Moshammer, B. Feuerstein, W. Schmitt, A. Dorn, C. D. Schröter, J. Ullrich, H. Rottke, C. Trump, M. Wittmann, G. Korn, K. Hoffmann, W. Sandner, Momentum distributions of Ne^{n+} ions created by an intense ultrashort laser pulse. Phys. Rev. Lett. **84**, 447 (2000)
54. R. Moshammer, J. Ullrich, B. Feuerstein, D. Fischer, A. Dorn, C.D. Schrter, J.R.C. Lpez-Urrutia, C. Hhr, H. Rottke, C. Trump, M. Wittmann, G. Korn, K. Hoffmann, W. Sandner, Strongly directed electron emission in non-sequential double ionization of ne by intense laser pulses. J. Phys. B: At. Mol. Opt. Phys. **36**(6), L113–L119 (2003)
55. A. Rudenko, T. Ergler, K. Zrost, B. Feuerstein, V.L.B. de Jesus, C.D. Schrter, R. Moshammer, J. Ullrich, From non-sequential to sequential strong-field multiple ionization: identification of pure and mixed reaction channels. J. Phys. B: At. Mol. Opt. Phys. **41**(8), 081006 (2008)
56. R. Dörner, V. Mergel, O. Jagutzki, L. Spielberger, J. Ullrich, R. Moshammer, H. Schmidt-Bcking, Cold target recoil ion momentum spectroscopy: a 'momentum microscope' to view atomic collision dynamics. Phys. Rep. **330**, 95–192 (2000)
57. A. Rudenko, K. Zrost, B. Feuerstein, V.L.B. de Jesus, C. D. Schröter, R. Moshammer, J. Ullrich, Correlated multielectron dynamics in ultrafast laser pulse interactions with atoms. Phys. Rev. Lett. **93**, 253001 (2004)
58. V.L.B. de Jesus, B. Feuerstein, K. Zrost, D. Fischer, A. Rudenko, F. Afaneh, C.D. Schröter, R. Moshammer, J. Ullrich, Atomic structure dependence of nonsequential double ionization of He, Ne and Ar in strong laser pulses. J. Phys. B: At. Mol. Opt. Phys. **37**(8), L161–L167 (2004)
59. A.S. Alnaser, D. Comtois, A.T. Hasan, D.M. Villeneuve, J. Kieffer, I.V. Litvinyuk, Strong-field non-sequential double ionization: wavelength dependence of ion momentum distributions for neon and argon. J. Phys. B: At. Mol. Opt. Phys. **41**(3), 031001 (2008)
60. X. Liu, H. Rottke, E. Eremina, W. Sandner, E. Goulielmakis, K.O. Keeffe, M. Lezius, F. Krausz, F. Lindner, M.G. Schtzel, G.G. Paulus, H. Walther, Nonsequential double ionization at the Single-Optical-Cycle limit. Phys. Rev. Lett. **93**, 263001 (2004)

61. S. Baker, J.S. Robinson, C.A. Haworth, H. Teng, R.A. Smith, C.C. Chiril, M. Lein, J.W.G. Tisch, J.P. Marangos, Probing proton dynamics in molecules on an attosecond time scale. Science **312**(5772), 424–427 (2006)
62. J.I. Steinfeld, *Molecules and Radiation: Introduction to Modern Molecular Spectroscopy* (MIT Press, Cambridge, 1978)
63. D. Tannor, *Introduction to Quantum Mechanics: A Time-Dependent Perspective*, U.S., illustrated edn. (University Science Books, Sausalito, 2007)
64. H. Lefebvre-Brion, R.W. Field, *The Spectra and Dynamics of Diatomic Molecules: Revised and Enlarged Edition*, 1st edn. (Academic Press, New York, 2004)
65. W. Demtrder, *Atoms, Molecules and Photons: An Introduction to Atomic- Molecular- and Quantum Physics* (Springer, Berlin, 2005)
66. P.B. Bransden, P.C. Joachain, *Physics of Atoms and Molecules*, 2nd edn. (Prentice Hall, Harlow, 2003)
67. M.J. Winter, *Chemical Bonding* (Oxford University Press, Oxford,1994)
68. P. Atkins, R. Friedman, *Molecular Quantum Mechanics*, 4th edn. (Oxford University Press, Oxford, 2005)
69. Bandrauk, *Molecules in Laser Fields*, 1st edn. (CRC Press, Boca Raton, 1993)
70. J. Franck, E.G. Dymond, Elementary processes of photochemical reactions. Trans. Faraday Soc. **21**, 536 (1926)
71. E.U. Condon, Nuclear motions associated with electron transitions in diatomic molecules. Phys. Rev. **32**, 858–872 (1928)
72. J.H. Posthumus, The dynamics of small molecules in intense laser fields. Rep. Prog. Phys. **67**(5), 623–665 (2004)
73. L.J. Frasinski, J.H. Posthumus, J. Plumridge, K. Codling, P.F. Taday, A.J. Langley, Manipulation of bond hardening in H_2^+ by chirping of intense femtosecond laser pulses. Phys. Rev. Lett. **83**, 3625–3628 (1999)
74. L.J. Frasinski, J. Plumridge, J.H. Posthumus, K. Codling, P.F. Taday, E.J. Divall, A.J. Langley, Counterintuitive alignment of H_2^+ in intense femtosecond laser fields. Phys. Rev. Lett. **86**, 2541–2544 (2001)
75. P.H. Bucksbaum, A. Zavriyev, H.G. Muller, D.W. Schumacher, Softening of the H_2^+ molecular bond in intense laser fields. Phys. Rev. Lett. **64**, 1883–1886 (1990)
76. A. Giusti-Suzor, F.H. Mies, Vibrational trapping and suppression of dissociation in intense laser fields. Phys. Rev. Lett. **68**, 3869–3872 (1992)
77. J.J. Larsen, Laser induced alignment of neutral molecules. Ph.D. thesis, Institute of Physics and Astronomy, University of Aarhus, Denmark (2000)
78. K. Codling, L.J. Frasinski, Coulomb explosion of simple molecules in intense laser fields. Contemp. Phys. **35**(4), 243 (1994)
79. M. Abu-samha, L.B. Madsen, Theory of strong-field ionization of aligned Co_2. Phys. Rev. A **80**(2), 023401 (2009)
80. X.M. Tong, Z.X. Zhao, C.D. Lin, Theory of molecular tunneling ionization. Phys.Rev. A **66**(3), 033402 (2002)
81. B.K. McFarland, J.P. Farrell, P.H. Bucksbaum, M. Guhr, High harmonic generation from multiple orbitals in N_2. Science **322**, 1232–1235 (2008)
82. W. Boutu, S. Haessler, H. Merdji, P. Breger, G. Waters, M. Stankiewicz, L.J. Frasinski, R. Taieb, J. Caillat, A. Maquet, P. Monchicourt, B. Carre, P. Salieres, Coherent control of attosecond emission from aligned molecules. Nat. Phys. **4**(7), 545–549 (2008)
83. X. Jia, W. Li, J. Liu, J. Chen, Alignment-dependent nonsequential double ionization of N_2 in intense laser fields: the role of different valence orbitals. Phys. Rev. A **80**, 053405 (2009)
84. O. Smirnova, Y. Mairesse, S. Patchkovskii, N. Dudovich, D. Villeneuve, P. Corkum, M.Y. Ivanov, High harmonic interferometry of multi-electron dynamics in molecules. Nature **460**(7258), 972–977 (2009)
85. D. Pavicic, K.F. Lee, D.M. Rayner, P.B. Corkum, D.M. Villeneuve, Direct measurement of the angular dependence of ionization for N_2, O_2, and Co_2 in intense laser fields. Phys. Rev. Lett. **98**(24), 243001 (2007)

86. G.A. Gallup, I.I. Fabrikant, Semiclassical complex-time method for tunneling ionization: molecular suppression and orientational dependence. Phys. Rev. A **81**(3), 033417 (2010)
87. S. Petretti, Y.V. Vanne, A. Saenz, A. Castro, P. Decleva, Alignment-dependent ionization of N_2, O_2, and Co_2 in intense laser fields. Phys. Rev. A **104**(22), 223001 (2010)
88. R. Murray, M. Spanner, S. Patchkovskii, M.Y. Ivanov, Tunnel ionization of molecules and orbital imaging. Phys. Rev. A **106**(17), 173001 (2011)
89. L. Torlina, M. Ivanov, Z.B. Walters, O. Smirnova, Time-dependent analytical R-matrix approach for strong-field dynamics. II. many-electron systems. Phys. Rev. A **86**, 043409 (2012)
90. S.M. Hankin, D.M. Villeneuve, P.B. Corkum, D.M. Rayner, Nonlinear ionization of organic molecules in high intensity laser fields. Phys. Rev. Lett. **84**, 5082 (2000)
91. S.M. Hankin, D.M. Villeneuve, P.B. Corkum, D.M. Rayner, Intense-field laser ionization rates in atoms and molecules. Phys. Rev. A **64**, 013405 (2001)
92. T. Koopmans, Über die zuordnung von wellenfunktionen und eigenwerten zu den einzelnen elektronen eines atoms. Physica **1**(16), 104–113 (1934)
93. C. Guo, M. Li, J.P. Nibarger, G.N. Gibson, Single and double ionization of diatomic molecules in strong laser fields. Phys. Rev. A **58**, R4271 (1998)
94. A. Rudenko, B. Feuerstein, K. Zrost, V.L.B. de Jesus, T. Ergler, C. Dimopoulou, C.D. Schrter, R. Moshammer, J. Ullrich, Fragmentation dynamics of molecular hydrogen in strong ultrashort laser pulses. J. Phys. B: At. Mol. Opt. Phys. **38**(5), 487–501 (2005)
95. E. Eremina, X. Liu, H. Rottke, W. Sandner, M.G. Schuetzel, A. Dreischuh, G.G. Paulus, H. Walther, R. Moshammer, J. Ullrich, Influence of molecular structure on double ionization of N_2 and O_2 by high intensity ultrashort laser pulses. Phys. Rev. Lett. **92**, 173001 (2004)
96. C. Guo, G.N. Gibson, Ellipticity effects on single and double ionization of diatomic molecules in strong laser fields. Phys. Rev. A **63**(4), 040701 (2001)
97. L. Pei, C. Guo, Nonsequential double ionization of triatomic molecules in strong laser fields. Phys. Rev. A **82**(2), 021401 (2010)
98. A.S. Alnaser, X.M. Tong, T. Osipov, S. Voss, C.M. Maharjan, P. Ranitovic, B. Ulrich, B. Shan, Z. Chang, C.D. Lin, C.L. Cocke, Routes to control of H_2 coulomb explosion in few-cycle laser pulses. Phys. Rev. Lett. **93**, 183202 (2004)
99. A.S. Alnaser, C.M. Maharjan, X.M. Tong, B. Ulrich, P. Ranitovic, B. Shan, Z. Chang, C.D. Lin, C.L. Cocke, I.V. Litvinyuk, Effects of orbital symmetries in dissociative ionization of molecules by few-cycle laser pulses. Phys. Rev. A **71**, 031403 (2005)
100. P. Dietrich, D.T. Strickland, M. Laberge, P.B. Corkum, Molecular reorientation during dissociative multiphoton ionization. Phys. Rev. A **47**, 2305 (1993)
101. X.M. Tong, Z.X. Zhao, A.S. Alnaser, S. Voss, C.L. Cocke, C.D. Lin, Post ionization alignment of the fragmentation of molecules in an ultrashort intense laser field. J. Phys. B: At. Mol. Opt. Phys. **38**(4), 333–341 (2005)
102. C. Ellert, P.B. Corkum, Disentangling molecular alignment and enhanced ionization in intense laser fields. Phys. Rev. A **59**, R3170 (1999)
103. S. Voss, A.S. Alnaser, X. Tong, C. Maharjan, P. Ranitovic, B. Ulrich, B. Shan, Z. Chang, C.D. Lin, C.L. Cocke, High resolution kinetic energy release spectra and angular distributions from double ionization of nitrogen and oxygen by short laser pulses. J. Phys. B: At. Mol. Opt. Phys. **37**(21), 4239–4257 (2004)
104. I. Litvinyuk, Alignment-dependent strong field ionization of molecules. Phys. Rev. Lett. **90**, 233003 (2003)
105. J. Liu, D.F. Ye, J. Chen, X. Liu, Complex dynamics of correlated electrons in molecular double ionization by an ultrashort intense laser pulse. Phys. Rev. Lett. **99**, 013003 (2007)
106. Y. Li, J. Chen, S.P. Yang, J. Liu, Alignment effect in nonsequential double ionization of diatomic molecules in strong laser fields. Phys. Rev. A **76**(2), 023401 (2007)
107. X. Liu, C. Wu, Z. Wu, Y. Liu, Y. Deng, Q. Gong, Recollision-induced dissociation and ionization of oxygen in few-cycle laser fields. Phys. Rev. A **83**(3), 035403 (2011)
108. J. Wu, H. Zeng, C. Guo, Polarization effects on nonsequential double ionization of molecular fragments in strong laser fields. Phys. Rev. A **75**, 043402 (2007)

109. C. Guo, M. Li, J.P. Nibarger, G.N. Gibson, Nonsequential double ionization of molecular fragments. Phys. Rev. A **61**(3), 033413 (2000)
110. Z. Wu, C. Wu, Q. Liang, S. Wang, M. Liu, Y. Deng, Q. Gong, Fragmentation dynamics of methane by few-cycle femtosecond laser pulses. J. Chem. Phys. **126**, 074311 (2007)
111. C. Wu, Z. Wu, Q. Liang, M. Liu, Y. Deng, Q. Gong, Ionization and dissociation of alkanes in few-cycle laser fields. Phys. Rev. A **75**, 043408 (2007)
112. F.A. Rajgara, M. Krishnamurthy, D. Mathur, Electron rescattering and the fragmentation dynamics of molecules in strong optical fields. Phys. Rev. A **68**(2), 023407 (2003)
113. V.R. Bhardwaj, D.M. Rayner, D.M. Villeneuve, P.B. Corkum, Quantum interference in double ionization and fragmentation of C_6H_6 in intense laser fields. Phys. Rev. Lett. **87**, 253003 (2001)
114. F.A. Rajgara, M. Krishnamurthy, D. Mathur, Electron rescattering and the dissociative ionization of alcohols in intense laser light. J. Chem. Phys. **119**(23), 12224 (2003)
115. V.R. Bhardwaj, P.B. Corkum, D.M. Rayner, Recollision during the high laser intensity ionization of C_{60}. Phys. Rev. Lett. **93**, 043001 (2004)
116. M. Murakami, M. Tanaka, T. Yatsuhashi, N. Nakashima, Enhancement of anthracene fragmentation by circularly polarized intense femtosecond laser pulse. J. Chem. Phys. **126**(10), 104304 (2007)
117. T. Yatsuhashi, N. Nakashima, Formation and fragmentation of quadruply charged molecular ions by intense femtosecond laser pulses. J. Phys. Chem. A **114**, 7445–7452 (2010)
118. A. Talebpour, A.D. Bandrauk, K. Vijayalakshmi, S.L. Chin, Dissociative ionization of benzene in intense ultra-fast laser pulses. J. Phys. B: At. Mol. Opt. Phys. **33**(21), 4615 (2000)
119. D. Mathur, A.K. Dharmadhikari, F.A. Rajgara, J.A. Dharmadhikari, Molecular symmetry effects in the ionization of CS_2 by intense, few-cycle laser pulses (2008), arXiv:0801.2046v1

Chapter 3
Lasers

The main tool for the experiments presented in this thesis are ultrashort, high power laser pulses. When focussed, the energy carried by such a light pulse is highly confined in both space and time, leading to peak intensities that surpass $10^{14}\,\mathrm{Wcm^{-2}}$. This makes it possible to induce highly nonlinear light matter interactions that can be used as probes for molecular dynamics. In this chapter, I will present the most important processes that underlie the production and characterisation of femtosecond (fs) laser pulses as used in the presented experiments.

The word *LASER* is an acronym for Light Amplification (through) Stimulated Emission (of) Radiation. The underlying process of stimulated emission was theoretically predicted by Einstein in 1916 [1] and experimentally demonstrated by Maiman in 1960 [2] with the invention of the laser. From a technological point of view, a laser consists of an optical resonator, a gain or laser medium and an energy pump source. Such an arrangement is called a laser *oscillator*. In the context of strong field physics, titanium doped with sapphire (Ti:S) [3] has become the most popular laser medium due to its high damage threshold, large emission bandwidth and an absorption band [4] that allows for efficient energy pumping through commercially available pump sources. Here, such a source usually is a diode pumped solid state laser emitting 532 nm radiation in continuous wave (CW) operation. The Ti:S oscillator has become the workhorse of strong field physics, producing ultrashort pulses at 800 nm with a duration down to 5 fs [5]. This is commonly achieved through a technique called Kerr-lens mode-locking (see 3.1.5 and 3.2.1) and careful dispersion control (see 3.1.2). Higher energies per pulse are now routinely achieved through laser amplifiers employing the *chirped pulse amplification* (CPA) technique (see 3.2.2) enabling the production of 30 fs pulses of several mJ at a few kHz repetition rate. Even shorter pulses have been produced through external pulse broadening and post compression techniques and novel nonlinear optical processes such as high harmonic generation (see 2.1.5).

M. Oppermann, *Resolving Strong Field Dynamics in Cation States of CO_2 via Optimised Molecular Alignment*, Springer Theses, DOI: 10.1007/978-3-319-05338-7_3,

3.1 Laser Fundamentals

This section presents the most important nonlinear processes driven by an ultrashort laser pulse that are required for describing the basic architecture of a femtosecond laser system.

3.1.1 *Mathematical Description*

A laser pulse is an electromagnetic wavepacket containing a continuous set of frequencies. In the following discussion I will neglect the magnetic field component of the laser field, because at the light intensities relevant to this thesis its amplitude remains sufficiently low to not play a role in light-matter interactions. This approximation is justified, because the coupling of the laser electric field to the molecular states is much stonger than that induced by the magnetic field component. Mathematically it is then possible to decompose the remaining electric field of a laser pulse into a (complex) envelope function $\tilde{E}(t)$ and a sinusoidal carrier with frequency ω_L [6] (see Fig. 3.1). This chapter is mostly concerned with the temporal features of the pulse, such that its spatial dependence is omitted in the following treatment:

$$E_L(t) = E(t)e^{i\omega_L t}e^{i\phi(t)} + c.c. \tag{3.1}$$

Here, $c.c.$ denotes the complex conjugate of the expression. I will omit the $c.c.$ in subsequent equations for convenience. Furthermore, the intensity of the laser pulse is related to its electric field as $I(t) = \frac{c_0 \epsilon_0 n}{2}|E_L(t)|^2$, where c_0 denotes the vacuum speed of light, $epsilon_0$ the vacuum permittivity and n the refractive index of the propagated medium. $\phi(t)$ denotes the *temporal phase* of the pulse, which contains its frequency versus time information. In the case of $\phi(t) = const. = \phi_0$, it corresponds to the offset between carrier and envelope called *carrier envelope phase* (CEP). The full exponent of the carrier function describes the temporal behaviour of the pulse's frequencies. Its time derivative therefore gives the instantaneous angular frequency ω:

$$\omega(t) = \omega_L + \frac{\partial \phi(t)}{\partial t} \tag{3.2}$$

According to the mathematical structure of the pulse, the envelope moves with the so-called *group velocity*:

$$v_g = \frac{\partial \omega}{\partial k} = \left(\frac{\partial k}{\partial \omega}\right)^{-1} \tag{3.3}$$

where $k = 2\pi/\lambda$ denotes the (angular) wavenumber. Its phasefronts and hence the carrier moves at the so-called *phase velocity*:

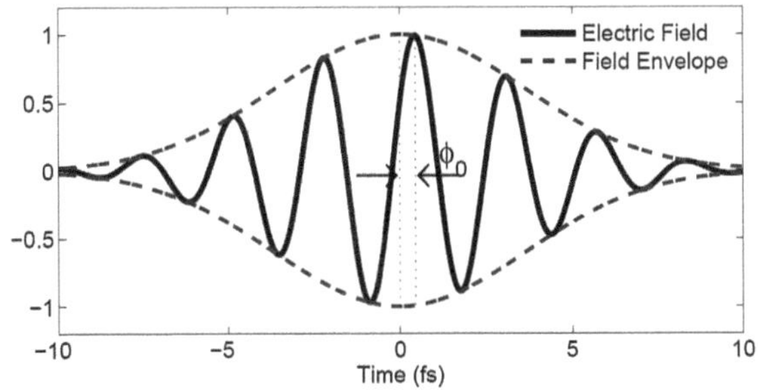

Fig. 3.1 The above figure illustrates the mathematical description of a laser pulse, with the carrier as the solid and the envelope as the *dashed line*. The picture also shows the phase off-set between carrier and envelope called carrier-envelope phase ϕ_0. Taken from [7], courtesy of Felix Frank

$$v_{ph} = \frac{\omega}{k} \tag{3.4}$$

Fourier transforming Eq. (3.1) leads to the representation of the pulse in the frequency domain, as shown in (3.5). The inverse transfrom (3.6) then leads back to the time domain:

$$\tilde{E}_L(\omega) = \int_{-\infty}^{\infty} E_L(t)e^{-i\omega t}dt \tag{3.5}$$

$$E_L(t) = (2\pi)^{-1} \int_{-\infty}^{\infty} \tilde{E}_L(\omega)e^{i\omega t}d\omega \tag{3.6}$$

where $S(\omega) = |\tilde{E}(\omega)|^2$ is the spectrum of the pulse. It is the above connection between the temporal and spectral characteristics of the pulse, which intertwines its duration τ and bandwidth $\Delta\omega$. This mutual dependence takes the form of a minimum duration bandwidth product [8]:

$$\tau\Delta\omega \geq 2\pi c_B \tag{3.7}$$

The numerical constant c_B is determined by the shape of the pulse envelope. For a Gaussian pulse, $c_B \approx 0.441$, whereas for a sech envelope one obtains $c_B \approx 0.315$ [8]. In general one has to note that for a pulse to be short, its spectrum has to be broad. A pulse, which fulfills Eq. (3.7) as an equality is called transform limited. Physically this means that all frequencies are present at every point in the time domain. In the frequency domain this means that the spectral phase function is constant.

3.1.2 Dispersion Effects

For a short laser pulse propagating through a dense medium, each of its spectral components will travel at a different phase velocity. This means that the pulse can broaden in the temporal domain. For a more detailed description, one may start with the dispersion relation from linear optics, which shows the dependence of the phase velocity on the refractive index $n(\omega)$ [8]:

$$k(\omega) = \frac{\omega}{v_{ph}(\omega)} = \frac{\omega}{c_0} n(\omega) \tag{3.8}$$

Above, c_0 denotes the vacuum speed of light. Note that in vacuum, $n(\omega) = 1$, such that $v_{ph} = v_g = c_0$. The behaviour of the spectral components of the pulse relative to each other are encoded in its spectral phase. From an experimental point of view, one wants to know how the pulse's envelope gets distorted after propagating a distance l in a dispersive medium. For achieving this, one may start with a transform limited pulse propagating in z-direction:

$$E(z, t) = \tilde{E}(z, t) e^{i\omega_L t} e^{i(k(\omega)z + \phi_0)} \tag{3.9}$$

The Fourier transform leaves the temporal phase invariant and hence one obtains $\varphi(z, \omega) = k(\omega)z + \phi_0$ for the spectral phase. As ϕ_0 is the same for every spectral component, it does not cause dispersion, hence the relative spectral phase shift for one component after propagation length $z = l$ is:

$$\Delta\varphi(\omega) = \frac{\omega n(\omega)}{c_0} l \tag{3.10}$$

The numerical value of the refractive index for each component can be obtained from the Sellmeier equation:

$$n^2(\lambda) = 1 + \sum_i \frac{B_i \lambda^2}{\lambda^2 - C_i} \tag{3.11}$$

where B_i and C_i are empirical constants, called the Sellmeier coefficients. In order to obtain the temporal envelope of the pulse after propagation through the medium, one would measure its spectrum before the medium and multiply each spectral amplitude by the corresponding phase from (3.10). The inverse transform of the resulting spectral envelope then gives the temporal one.

The dispersion of the laser pulse is characterized by the evolution of its spectral phase. For a more detailed understanding, it is helpful to write it as a Taylor expansion around the central frequency:

$$\varphi(\omega) = \phi_0 + \sum_{m=1}^{\infty} \frac{1}{m!} \frac{d^m \varphi(\omega)}{d\omega^m}\bigg|_{\omega=\omega_L} (\omega - \omega_L)^m$$
$$\equiv \phi_0 + \sum_{m=1}^{\infty} \frac{1}{m!} \varphi^{(m)}(\omega_L)(\omega - \omega_L)^m \tag{3.12}$$

Note that the first term still discribes the CEP, whereas the dispersion induced by the other terms in the expansion depends on the coefficients given by the derivatives $\varphi^{(m)}(\omega_L)$. Whether the *mth* phase dispersion leads to a substantial modification of the pulse is determined by the following scaling rule for the coefficients [6]:

$$\varphi^{(m)}(\omega_L) \geq \tau^m \tag{3.13}$$

where τ denotes the duration of the pulse. The effect of higher order dispersion onto the temporal duration of the pulse can be investigated by inverse transforming the frequency representation. Here, one usually starts by examining the impact of the individual terms in the Taylor expansion. Here, the mth-order term is usually referred to as the mth-*orderphase*. The *first-order phase* results in a shift in time, due to the Fourier Transform Shift Theorem:

$$(2\pi)^{-1} \int_{-\infty}^{\infty} \tilde{E}_L(\omega) e^{i\varphi^{(1)}(\omega_L)(\omega-\omega_L)} e^{i\omega t} d\omega = E_L(t + \varphi^{(1)}(\omega_L)) e^{-i\varphi^{(1)}(\omega_L)\omega_L} \tag{3.14}$$

Note that the additional phase factor in the exponent will only induce a CEP shift. Using Eq. (3.10) for the spectral phase accumulated in the dispersive medium, the shift in time corresponds to:

$$\varphi^{(1)}(\omega_L) = \frac{\partial k(\omega)}{\partial \omega}\bigg|_{\omega=\omega_L} l = \frac{l}{v_g} \tag{3.15}$$

which is the time the pulse takes to traverse the medium.

With a similar approach, one may also show that a *second order phase* leads to a linear temporal phase [10]. This means that the carrier frequency is varying linearly with time under the pulse envelope, which is called a linear *chirp*. If $\varphi^{(2)} > 0$, the chirp is positive and the frequency is increasing with time (and vice versa for negative chirp). Typically, the spectral phase will contain also higher-order phases, which leads to significant distortions of the pulse in the temporal domain. For ultrashort laser pulses, the acquired phase may become so complicated that Eq. (3.12) cannot be truncated after the first few terms. In practice one therefore often makes use of the numerical treatment illustrated previously.

The knowledge of the acquired spectral phase through a medium is important for being able to compensate for it. Most transparent materials add a postive second order phase to a pulse centred at 800 nm, so one needs methods to impose negative

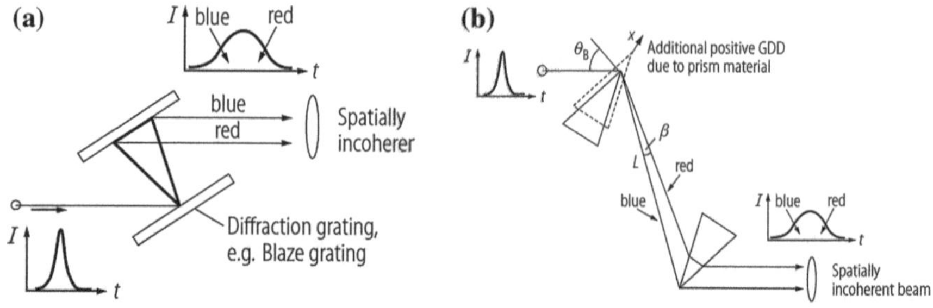

Fig. 3.2 Schematic representation of a grating (**a**) and prism (**b**) based configuration for manipulating the spectral phase of a laser pulse. In such setups, the gratings or prisms are used to disperse and hence separate the spectral components and induce a spectral phase by different individual optical pathlengths. Negative dispersion means that the short wavelength components travel a shorter pathway than the longer ones. (Reprinted from [9], with permission from Springer Science+Business Media)

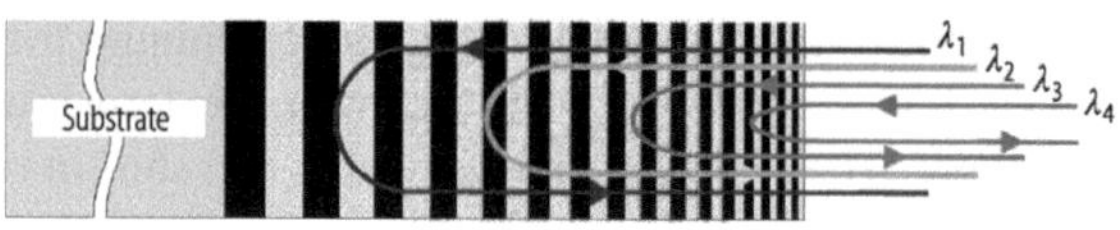

Fig. 3.3 Chirped mirrors are dielectric multilayer mirrors. Due to a modulation of the period of the multilayer stacks, longer wavelengths travel further into the mirror before being reflected. This is due to the constructive interference condition for the partial waves of one spectral component reflected at the interfaces between low and high refractive index layers [6]. As a result, longer wavelengths travel a longer optical path upon reflection than shorter ones and hence a negative dispersion is induced. (Reprinted from [9], with permission from Springer Science+Business Media)

dispersion. This can be done by pulse compressor setups, either using prisms or gratings. The underlying basic principle is presented in Fig. 3.2. The most complete and therefore most popular method of creating transform limited, short pulses with $\tau < 10\,\text{fs}$ is the use of chirped mirrors. They are described in Fig. 3.3.

3.1.3 Nonlinear Optics

From a classical perspective, the interaction of light and matter can be modeled by the interaction of the electromagnetic light field with a collection of dipoles representing the atoms or molecules in the matter system. *Linear optics* is the regime of light matter interactions, where the response of the system to the incoming electric field is independent of its intensity. However, upon increasing the field intensity, the matter system itself is perturbed by the field, which changes its response. In this *perturbative regime*, the electric laser field acts as a small perturbation onto the bound electrons in the Coulomb field of the atoms or molecules in the matter system. In the mathematical framework of lowest order perturbation theory (LOPT), the induced dipole moment per unit volume in the medium, or *polarisation*, can then be expressed as a Taylor expansion in the electric field strength of the incoming radiation [11]:

$$P = \epsilon_0 \left(\chi^{(1)} E + \chi^{(2)} E^2 + \chi^{(3)} E^3 + \cdots \right) \tag{3.16}$$

where ϵ_0 is the vacuum permeability and $\chi^{(n)}$ called the *nth* order susceptibility of the medium. Note that in the above expression the response is assumed to be instantaneous and isotropic. As the polarisation contains nonlinear terms with respect to the field strength, this regime is usually referred to as *nonlinear optics*. However, with the onset of field ionisation at intensities around 10^{14} Wcm^{-2} non-perturbative treatments are required, which is discussed in detail in Chap. 2.

Nonlinear optical effects in the perturbative regime are now routinely accessed and controlled in laser technology. In the context of the laser systems employed for this thesis, the second order process ($\chi^{(2)}$) *optical parametric amplification* and the third order ($\chi^{(3)}$) *optical Kerr effect* are of particular importance. They are briefly presented in the following sections.

3.1.4 Optical Parametric Amplification

Optical parametric processes comprise a set of phenomena where two light waves interact in a medium with nonzero $\chi^{(2)}$ [12]. Typically this results in the generation of new frequencies. With interacting waves of frequencies ω_1 and ω_2, generating ω_3, the process has to obey energy conservation. In the photon picture, this can be formulated as:

$$\hbar\omega_3 = \hbar\omega_1 + \hbar\omega_2 \Leftrightarrow \omega_3 = \omega_1 + \omega_2 \tag{3.17}$$

The generated frequency thus depends on the interacting frequencies. Momentum conservation then requires:

$$\hbar\mathbf{k}_3 = \hbar\mathbf{k}_1 + \hbar\mathbf{k}_2 \Leftrightarrow \mathbf{k}_3 = \mathbf{k}_1 + \mathbf{k}_2 \tag{3.18}$$

where $\mathbf{k}_i$ refers to the wave vector of the associated light wave. This condition is usually referred to as *phase-matching* and needs to be fulfilled for the process to be efficient. This is typically achieved via tuning the orientation of the nonlinear crystal's symmetry axes with respect to the propagation of the interacting beams.

In *sum-frequency generation* or *upconversion*, the sum of the interacting frequencies is generated, as suggested in (3.17). If $\omega_1 = \omega_2$, this is called *second harmonic generation*. However, assuming $\omega_1 < \omega_2$, the frequency difference may be generated as well, which requires $\omega_2 = \omega_3 + \omega_1$ due to energy conservation. In terms of photon numbers during the process, this means that a photon of frequency ω_2 is converted into two photons, one of frequency ω_1 and and another of ω_3 [12]. The wave with ω_1 is thus amplified at the expense of ω_2. This process is called *optical parametric*

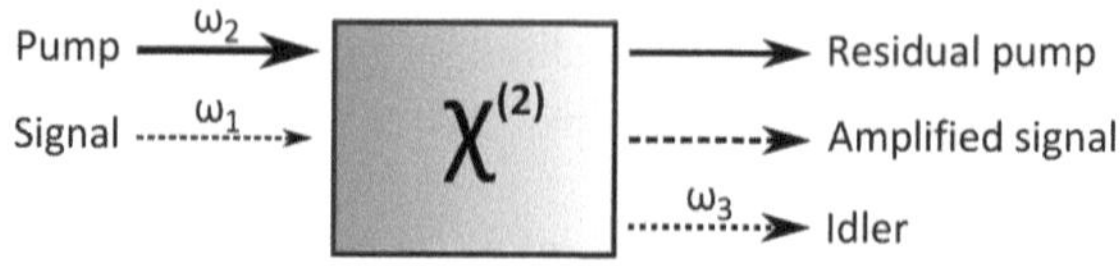

Fig. 3.4 Schematic representation of the optical parametric amplication process explained in the text

amplification (OPA) and is schematically illustrated in Fig. 3.4. In this context, ω_1 is referred to as the *seed* or *signal*, ω_2 the *pump* and ω_3 the *idler*. The amplification process requires a high photon flux of the pump, while the seed may be of much lower intensity in comparison. In OPA, energy is directly transferred from the pump to signal and idler, such that no energy is stored in the medium. This is different in amplification through population inversion.

3.1.5 Optical Kerr Effect

At sufficiently high intensity, the polarisation of a medium with nonzero $\chi^{(3)}$ results in an intensity dependent index of refraction [12]:

$$n(\mathbf{r}, t) = n_0 + n_2 I(\mathbf{r}, t) \tag{3.19}$$

where n_0 is the index of refraction from linear optics as introduced in Eq. (3.8) and n_2 the nonlinear index, determining the strength of the intensity dependence. Here, the laser pulse intensity $I(\mathbf{r}, t)$ is employed as a function of space and time. This phenomenon is called the optical Kerr effect. Its spatial dependence results in *self-focussing* of the laser beam, while its temporal dependence leads to the creation of new frequencies in the pulse in a process called *self phase modulation* (SPM). As $\chi^{(3)}$ does not vanish for centro-symmetric materials such as fused silica or gases, these effects are of prime importance when dealing with the propagation of intense laser pulses. Here, self-focussing may create hot spots in the laser beam that can damage optics, while SPM typically alters the temporal properties of the laser pulse. However, the control of both effects has lead to significant advances in laser pulse production, as will be presented in the next section.

Self-focussing

Assuming a Gaussian spatial intensity distribution for the laser pulse, the total index of refraction will have the same spatial dependence. This is illustrated Fig. 3.5. As a result, the medium acts as an intensity dependent Gaussian lens focussing the high intensity parts of the laser pulse. As this process is induced by the pulse itself, it is often referred to as self-focussing. Note that this effect increases upon propagation as the focussing further increases the peak intensity and thus the induced lensing effect. The focussing therefore continues either until the end of the medium or when the diffraction limit is reached.

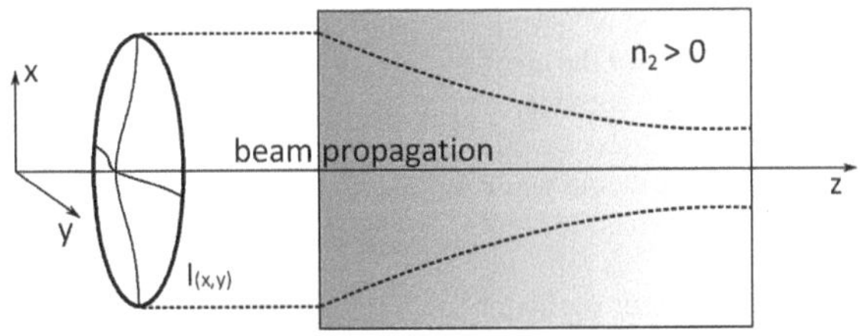

Fig. 3.5 Schematic illustration of the self-focussing process explained in the text. (Adapted from [12], with permission from Springer Science+Business Media)

Self phase modulation

Neglecting the spatial dependence of the laser pulse, one may include the now time dependent refractive index in the dispersion relation (3.8) and hence in the temporal phase of the pulse:

$$\phi(t, z) = \phi_0 + \frac{\omega_L}{c_0}(n_0 + n_2 I(t))z = \phi^{linear} + \frac{\omega_L n_2 z}{c_0} I(t) \tag{3.20}$$

The additional phase term resulting from the nonlinear refractive index leads to a spectral broadening upon propagation through the medium. This can be illustrated by assuming that the intensity variation is approximately linear at the edges of the pulse, such that:

$$\phi^{nonlin}(t) \propto n_2 \frac{dI(t)}{dt} t \tag{3.21}$$

The Fourier Transform Shift Theorem for a linear phase term has already been used in Eq. (3.14) from the spectral to the temporal domain. For the inverse transform, the shift has the opposite sign of the phase term, resulting from the definition of the transform in (3.5). Hence, in the case of the rising edge, $\frac{dI(t)}{dt} > 0$, the spectrum of the pulse is shifted to higher frequencies (blue shift), and to longer (red shift) for the falling edge. This means that the spectrum is broadened in both directions.

3.2 Laser Pulse Production

Pulsed laser operation offers several benefits for research in the optical sciences compared to narrow frequency continuous wave (CW) emission. At the same average power, the temporal confinement of the pulse energy leads to higher peak intensities, while the shorter interaction time with a target medium typically decreases damage caused by the interaction with the light. In the context of molecular physics however, ultrashort pulses have enabled time resolved molecular dynamics measurements. For this, the pulse duration has to be shorter than the duration of the molecular process. Technological advances for producing ever shorter pulse durations have thus been at the heart of both strong laser field physics and molecular dynamics research. Figure 3.6 shows the history of pulse production over the past three decades. In

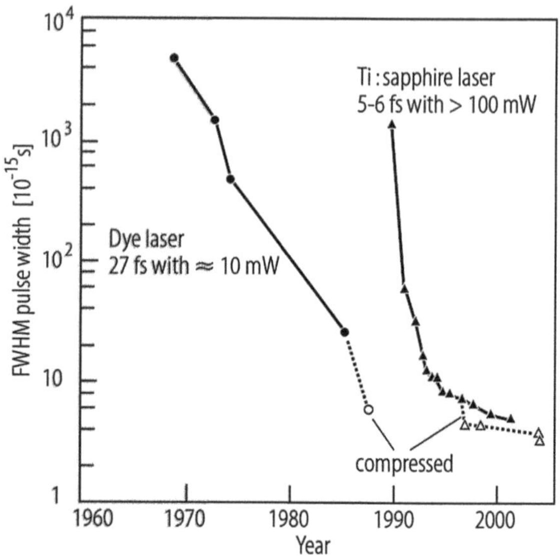

Fig. 3.6 Shortest achieved pulse duration over the course of the last three decades. *Circles* denote lasers based on dyes as gain media while triangles denote Ti:S technology. *Filled symbols* refer to pulses from a laser oscillator while *open symbols* denote durations obtained through external pulse compression techniques. (Reprinted from [9], with permission from Springer Science+Business Media)

the 1990s, Ti:S began to dominate ultrashort laser technology due to its large gain and emission bandwidth, enabling the production of spectrally broader and thus shorter duration pulses. Here, the first Ti:S oscillator was built in 1987 [3]. Kerr-Lens Mode-locking has become the most common pulse production technique for current Ti:S laser technology and is explained in Sect. 3.2.1. Currently, the shortest laser pulse obtained directly from an oscillator has been measured around 5 fs [5]. However, for obtaining such pulses with energies above 100 μJ, pulse amplification (see Sect. 3.2.2) followed by external spectral broadening and pulse compression has to be employed. This has been achieved via self-phase modulation in gas filled hollow core fibers [13] or filaments [14] followed by chirped mirror compressor setups. In the past decade, High Harmonic Generation (see Sect. 2.1.5) has become a reliable source of broadband pulses in the Extreme Ultraviolet spectral regime pushing the record of the shortest laser pulse to 80 attoseconds [15].

3.2.1 Mode-Locking

The principle of mode-locking (ML) is the constructive interference of the longitudinal modes supported by a laser cavity [12]. This leads to an intensity localisation in the temporal domain, such that the energy of all modes is confined to a pulse of duration τ. Longitudinal modes are standing waves between the cavity end mirrors. For a light wavelength λ and a cavity length l, the possible modes therefore have to obey the following relation:

$$q\lambda = 2l \tag{3.22}$$

where q is a positive integer. If all the modes oscillate at a constant phase with respect to each other, they may constructively interfere. The repetition rate R of the pulse train is then given by the frequency spacing between two adjacent modes $\Delta\nu$ and related to the round trip time T of a pulse in the cavity:

$$R = \Delta\nu = 1/T = \frac{c}{2l} \tag{3.23}$$

Note however that the number of modes N participating in the lasing process are restricted by the gain profile of the laser medium and $\Delta\nu$. Since $\Delta\nu N$ corresponds to the spectral bandwidth of the pulse, one may use the minimum duration bandwidth product to estimate the pulse duration as $\tau \approx T/N$. One can then relate the power of a single pulse P_p to the average power of the pulsed emission P_a [6]:

$$P_p = \frac{E}{\tau} \approx N\frac{E}{T} = NP_a \tag{3.24}$$

Here, E refers to the energy of a single pulse. By making one of the cavity end mirrors partially transparent, a train of laser pulses is thus coupled out of the cavity. The locking of the modes is enabled, when the loss of the laser cavity is modulated such that sufficient gain for lasing is only achieved in pulsed operation. Active mode-locking techniques thus rely on a so-called intracavity loss modulator that is externally driven at the repetition rate of the cavity. Passive mode-locking on the other hand relies on an intracavity saturable absorber that provides a significantly smaller loss for high intensity pulsed operation compared to low intensity CW operation of the cavity. In the latter case, ML is typically initiated by a small fluctuation in the CW amplitude. The corresponding intensity spike will acquire more gain than the low intensity CW light and will continue to do so with every round trip. The gain is thus depleted by the building laser pulse at the expense of the CW background until the entire light energy is contained in the pulse.

ML was first achieved in 1964 via an active technique in a Helium-Neon gas laser [16], while passive ML was first demonstrated in a Nd:Glass laser in 1966 [17]. Compared to active ML, passive ML usually provides a much faster loss modulation leading to much shorter pulse durations [9]. For this, passive techniques are now dominating ultrafast oscillator technology. Here, Kerr Lens Mode-locking (KLM) in Ti:S lasers [18] has become the most common arrangement in strong laser field science.

In KLM, the intracavity saturable absorber is an aperture smaller than the beam waist of the CW cavity mode. In pulsed operation however, the temporal confinement of the cavity radiation leads to a higher peak intensity and thus induces self-focussing in the laser medium. The reduced beam waist of the pulsed cavity mode then passes the aperture without additional losses. This process is illustrated schematically in Fig. 3.7. One usually distinguishes between hard and soft apertures. In the former case, a physical slit or pinhole is employed as suggested in the figure. However, in the case of a laser pumped oscillator, the pump beam waist may be adjusted such that the

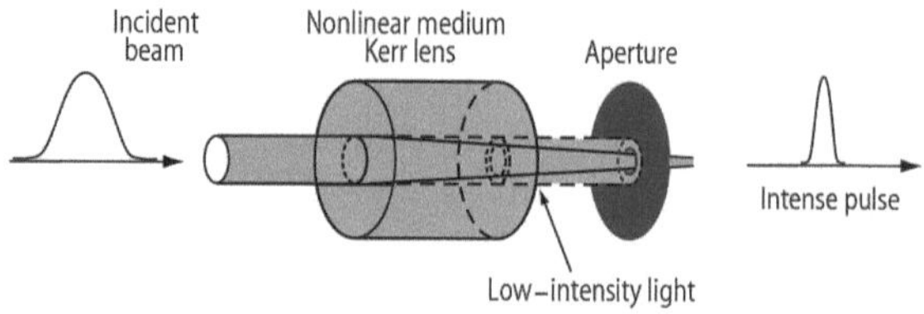

Fig. 3.7 Schematic illustration of Kerr Lens Mode-locking. A low intensity beam is clipped by the hard aperture, while a high intensity, short pulse beam induces self-focussing or a Kerr-lens in the nonlinear laser medium. The lensing effect reduces the beam waist of the high intensity pulse and thus its losses at the aperture. This favours pulsed over CW operation of the cavity. (Reprinted from [9], with permission from Springer Science+Business Media)

CW mode of the cavity is larger than the pumped cross section at the laser medium. Efficient gain is thus only possible in ML operation such that the self-focussed pulsed mode fully overlaps with the pump beam at the laser medium. In this case the pump beam waist provides a soft aperture for KLM.

Self-phase modulation and positive dispersion in the Ti:S crystal however limit the pulse length that can be obtained from an oscillator [6]. For this dispersion control elements have to be employed for compensation. Here, prism pairs or chirped mirrors are typical solutions.

3.2.2 *Chirped Pulse Amplification*

Pulse energies from oscillators are usually limited to the nJ level. For achieving high peak powers one thus needs to amplify the pulse. In the case of a Ti:S oscillator this can be achieved by passing the laser pulse through another Ti:S crystal with induced population inversion. However, due to self-focussing the amplified pulse intensity quickly reaches the damage threshold of the crystal, which makes it impossible to obtain pulse energies at the milliJoule level. To avoid such destructive propagation effects, the pulse may be stretched in time prior to amplification, such that the peak intensity of the amplified pulse remains below the damage threshold of the laser optics. This is the central idea behind the amplification technique called *Chirped Pulse Amplification* (CPA), developed in 1985 [19]. Here, the femtosecond pulse from the oscillator is chirped to ps duration via a pulse stretcher. This ensures that the pulse may be amplified to the mJ level in a Ti:S amplifier without inducing damage. Afterwards the initial chirp and dispersion introduced by the amplification stage maybe compensated for by a pulse compressor, recovering a transform limited pulse. A schematic illustration of a typical CPA arrangement is displayed in Fig. 3.8.

The gain achieved during amplification depends on the amplification medium, the initial pulse and pump power. Ti:S is very suitable for high power amplification due to a high damage threshold ($\approx$8–10 $\mathrm{J cm^{-2}}$) and high saturation fluence ($\approx$0.9 $\mathrm{J cm^{-2}}$) [4]. The latter denotes the maximum possible energy that can be extracted from the

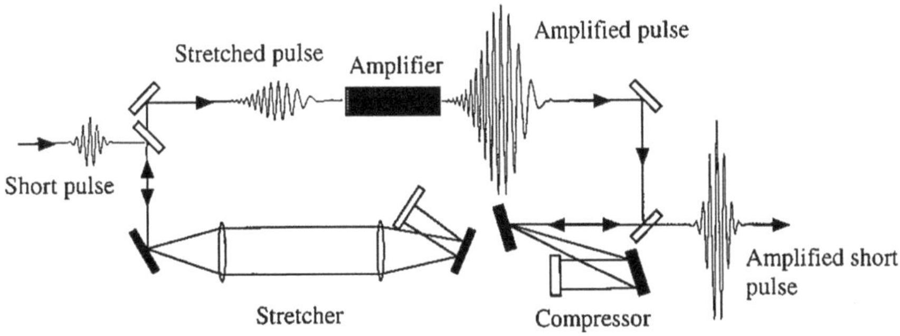

Fig. 3.8 Schematic illustration of a CPA setup. Note that grating based stretcher and compressor are utilised, where the *filled rectangles* denote gratings and *open rectangles* mirrors. (Reprinted with permission from [4]. Copyright (1998), AIP Publishing LLC)

medium per unit area [6]. The pump power is typically limited by the damage threshold and thermal conductivity of the gain medium. For this, the repetition rate of a mJ level Ti:S amplifier does usually not exceed a few kHz and requires external cooling of the amplifying medium. The obtained gain increases exponentially with increasing input pulse energy. This implies however that for an amplification over several orders of magnitude, several passes of the pulse through the amplifying medium are required. In this case, the frequency-dependent gain profile of the amplifying medium narrows the pulse spectrum with every pass, as spectral regions at the center of the profile are amplified most strongly. This effect is called *gain narrowing* and limits the spectral bandwidth of the output of a Ti:S amplifier to $\approx$47 nm [4].

3.3 Pulse Length Diagnostics

Measuring the duration of femtosecond laser pulses is typically not straight forward as the response time of electronic detectors is on the order of many picoseconds. For this all-optical methods have to be employed. Usually, accurately characterised reference pulses are not available so femtosecond pulse measurements normally rely on techniques involving the correlation of the unknown pulse with its replica. The simplest realisation of such an approach is intensity autocorrelation. However, this technique does not allow for the full reconstruction of the laser pulse and more advanced methods are required for achieving this. In this Ph.D. project Spectral Phase Interferometry (for) Direct Electric Field Reconstruction (SPIDER) was the only such technique employed. For this, another very popular pulse characterisation technique - Frequency Resolved Optical Gating (Frog) [20, 21]—is only briefly mentioned this thesis.

3.3.1 Autocorrelation and FROG

In an intensity autocorrelation, the yield of a nonlinear optical process is recorded as a function of delay τ between the pulse to be measured and its replica. It was first employed in 1967 [22, 23]. The intensity autocorrelation function $I_{AC}(\tau)$ is a measure of the pulse's intensity $I(t)$ in the time domain:

$$I_{AC}(\tau) = \int_{-\infty}^{\infty} I(t)I(t-\tau)dt \tag{3.25}$$

If the nonlinear optical signal is only produced when the pulses overlap in time and space, the full width of half maximum (FWHM) of $I_{AC}(\tau)$ is directly related to the pulse length in time. However, this measurement does not yield information about the temporal phase of the pulse and hence its envelope has to be guessed for extracting the pulse length $\Delta\tau$. For a Gaussian envelope the relationship results in $\Delta\tau = \Delta\tau_{AC}/1.41$. In practice the pulses are usually split via a beam splitter and delayed with respect to each other via a difference in optical path length. Typically, second harmonic generation (SHG) in a β barium borate (BBO) or lithium triborate (LBO) crystal is employed as the nonlinear optical process. The pulses are focussed onto the crystal and overlapped in time and space at the focus. The SHG signal is then recorded as a function of pulse delay.

FROG was developed [20, 21] to overcome the limitations of an autocorrelation measurement and retrieve the pulse's phase information as well. The FROG technique is essentially a spectrally resolved intensity autocorrelation, where the recorded FROG signal is given by:

$$I_{FROG}(\omega,\tau) = \left| \int_{-\infty}^{\infty} E(t)g(t-\tau)e^{-i\omega t}dt \right|^2 \tag{3.26}$$

where $E(t)$ is the electric field of the pulse and $g(t-\tau)$ a gate function that represents a nonlinear optical effect that depends on the pulse overlap. There are several different FROG geometries resulting from the range of nonlinear optical effects that can provide a suitable gate function. One of the most popular geometries uses second harmonic generation analoguously to the discussion of intensity autocorrelation above.

3.3.2 General SPIDER Technique

For a complete reconstruction of the laser pulse, both amplitude and phase have to be measured. Besides FROG, another technique that facilitates this is SPIDER

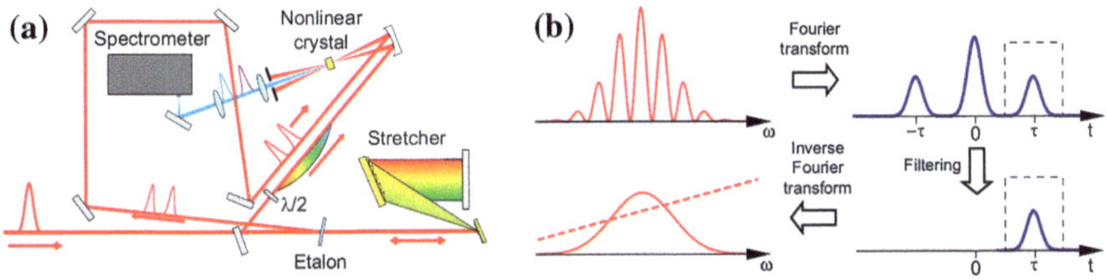

Fig. 3.9 **a** Schematic setup associated with the generalised SPIDER technique for pulse length measurements. **b** Depiction of the phase reconstruction procedure from an interferogram. (Reprinted from [25], with permission from Optical Society of America)

[24]. It is based on extracting phase information from the beating frequency of the interference between the laser pulse and a spectrally sheared replica. The associated setup is illustrated in part (a) of Fig. 3.9. The laser pulse is split into three copies—two low and one high energy replica. The two low energy pulses are temporally delayed by a time τ, which is typically on the order of a few picoseconds. They are called the *test pulses*. Both, splitting and delay, can be achieved via the use of an etalon for example. The high energy replica is called *ancilla* and is stretched in time via a linear chirp. This can be achieved with a block of glass or a grating stretcher. The spectral shear between the test pulses is now achieved by upconversion with different spectral parts of the ancilla. When overlapped with the ancilla in time and space in a nonlinear medium with nonzero $\chi^{(2)}$, each test pulse interacts with a different spectral slice of the ancilla. The frequency of the first test pulse is thus upconverted to $\omega_a = \omega + \omega_0$, while for the delayed second pulse $\omega_b = \omega + \omega_0 + \Omega$ is obtained. Here, Ω is the spectral shear between the pulses. The interferogram of the two upconverted test pulses $S(\omega)$ can then be written as:

$$S(\omega) = |E(\omega_a) + E(\omega_b)e^{i\omega\tau}|^2 \tag{3.27}$$

$$= |E(\omega_a)|^2 + |E(\omega_b)|^2 + 2|E(\omega_a)||E(\omega_b)| \cos\left(\Delta\phi(\omega) + \omega\tau\right) \tag{3.28}$$

Here, $E(\omega)$ is the spectrum of the original test pulses and $\Delta\phi(\omega) = \left(\phi(\omega_b) - \phi(\omega_a)\right)$ the phase difference of the upconverted pulses. The phase $\phi(\omega)$ of the test pulse is thus contained in the oscillations of the cosine term and can be extracted via Fourier-transform spectral interferometry [26], which is illustrated in part (b) of Fig. 3.9. For this, the spectrogram is Fourier transformed into the time domain where it consists of three peaks; one centred at $t = 0$ and two identical peaks at $t = \pm\tau$. The latter contain the phase information from the cosine term. Via a suitable filter function, one of the side peaks is isolated and inverse Fourier transformed back to the spectral domain to obtain $\Delta\phi(\omega) + \omega\tau$. The linear term can be obtained separately by applying the same procedure to the unsheared test pulses. In this way, $\Delta\phi(\omega)$ is extracted. The pulse's phase can then be reconstructed from this phase difference via an algebraic procedure often referred to as *phase concatenation*. Here, one starts with an arbitrarily chosen constant for the phase offset $\phi(\omega_0) = \phi_0$ and obtains the

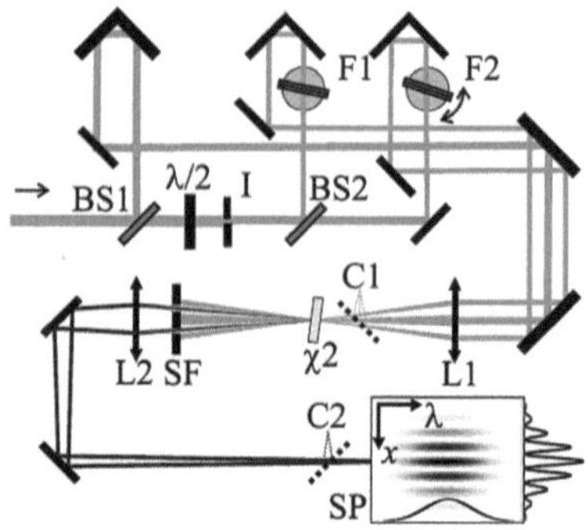

Fig. 3.10 Typical SEA-F-SPIDER setup. BS refers to beamsplitter while *F1* and *F2* denote the filters for producing the sheared ancillas. Note that in the SEA-F-SPIDER apparatus at Imperial College there is no half-wave plate in the ancilla beam path. *C1* and *C2* are flip mirrors for further beam characterisation, while *L1* and *L2* refer to imaging lenses. (Reprinted from [27], with permission from Optical Society of America)

phase function values at spectral frequencies spaced by the shear Ω:

$$\phi(\omega_n) = \sum_{i=0}^{n-1} \Delta\phi(\omega_i) \tag{3.29}$$

Here, $\omega_i = \omega_0 + i\Omega$, where i is a positive integer. For the described procedure to hold, one has to ensure that the stretching of the ancilla is sufficiently large such that the spectral slices for the upconversion process are quasi monocromatic over the duration of the test pulses. Furthermore the delay between the test pulses has to be long enough such that the side peaks from the Fourier transform are fully separated from the central peak.

3.3.3 SPIDER Apparatus

The SPIDER apparatus used for characterising laser pulses centred at 800 nm is commercially available (APE). The pulse is split via 100 µm thick etalon. The ancilla is stetched via a grating compressor with 830 lines/mm gratings. The stretcher design limits this apparatus to the measurement of pulse lengths between 20–80 fs. The nonlinear crystal used for the upconversion process is a type II BBO of 100 µm thickness. For this, the polarisation of the test pulses is rotated by 90° via a periscope before the crystal. The interference of the upconverted test pulses is recorded via a 1-dimensional spectrometer with 2048 pixels and a spectral range of 700–940 nm. Complete reconstruction of the measured pulse is facilitated via a commercial software.

3.3.4 NIR SEA-F-SPIDER Apparatus

For the characterisation of femtosecond laser pulses in the wavelength range from 1.1 to 2.2 μm, Tobias Witting has developed and built a SPIDER apparatus for the Laser Consortium at Imperial College London [28]. It differs from the previously described SPIDER in two aspects. Firstly, a home-built 2-dimensional spectrometer is used for recording the spectrogram. This allows for a spatio-temporal characterisation of the laser pulse. This approach has been coined *spatially encodeded arrangement* (SEA)—SPIDER. Secondly, the upconversion is achieved via two spectrally sheared and temporally delayed ancilla pulses and one test pulse. For this, the ancilla pulses are obtained directly from two pulse replicas via identical narrow band filters [27]. The shear is achieved by tuning the angle of one of the filters with respect to the laser pulse propagation. The use of filters then makes the apparatus a SEA-F-SPIDER. The schematic setup is shown in Fig. 3.10. Here, pulse splitting is achieved via beam splitters and upconversion done in a 50 μm thick type I BBO crystal.

3.4 The Imperial College CPA System

The Imperial College CPA laser is commercially available and consists of a Ti:S oscillator (Griffin, KM Labs) and a double stage amplifier (Red Dragon, KM Labs). The system delivers 30 fs pulses centered at 800 nm with a spectral bandwidth of $\Delta\omega \approx 45$ nm. Compressed pulse energies around 8 mJ are routinely generated at a 1 kHz repetition rate. The main components of the CPA system are described below.

3.4.1 Oscillator

The Ti:S oscillator (see Fig. 3.11) is pumped by a diode pumped Nd:YAG CW laser (Verdi, Coherent), delivering 5.2 W at 532 nm. Pulsed operation of the oscillator is achieved through Kerr-lens mode-locking with the pump beam as a soft aperture. Laser pulse dispersion during a cavity round trip is controlled via two motorised fused silica prisms. Their positioning is adjusted for optimising the center wavelength and bandwidth of the output pulses. When optimised for mode-locked operation, the oscillator provides a CW power of 300 to 400 mW. Moving one of the prisms typically initiates mode-locked with an increase in output power to 450 to 550 mW. Typically, larger differences between CW and mode-locked output power correspond to a more stable mode-locked operation. Ideally, the oscillator then delivers 20 fs pulses centred at 790 nm with a bandwidth of $\Delta\omega > 50$ nm and a Gaussian TEM_{00} mode. The laser operates at a repetition rate of approximately 89 MHz corresponding to a cavity roundtrip time of about 11.3 ns. This implies a pulse energy of 5 to 6 nJ. On a day to day basis the output spectrum is checked with a spectrometer. The repetition rate of

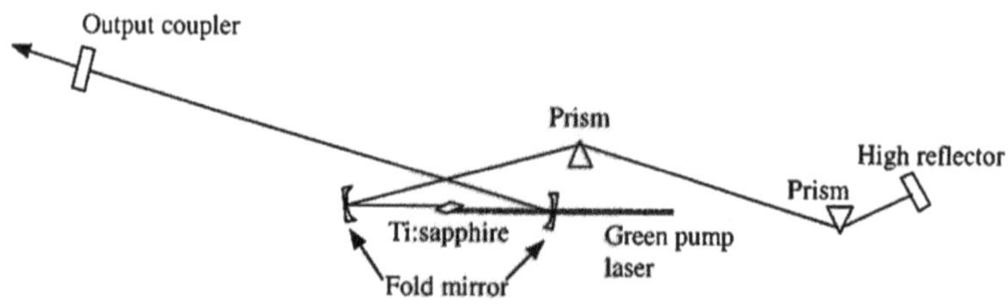

Fig. 3.11 Schematic illustration of the design of the oscillator in the Imperial College CPA system. Note that the folding mirrors act as focussing and collimating optics, respectively, while the focussing lens of the pump is not displayed. (Reprinted with permission from [4]. Copyright (1998), AIP Publishing LLC)

the oscillator is measured with a fast photodiode and the resulting electronic signal is used for triggering the components of the amplifier.

Optimisation of the oscillator output is achieved by adjusting the cavity end mirrors in conjunction with tuning the position of the Ti:S crystal, one of the cavity focussing mirrors and the pump focussing lens. Achieving ideal operating conditions is usually rather cumbersome as several laser output properties change upon adjusting a single optical component. However, experience has shown that the most crucial properties are mode-locked stability, centre wavelength and output mode. Through gain narrowing during amplification, the pulse bandwidth is typically limited to $\Delta\omega \approx 45\,\mathrm{nm}$ and the gain driven into saturation. This implies that bandwidths exceeding 50 nm or large mode-locked powers are not necessary. The center wavelength however has to be matched to the gain profile of the laser amplifier which was found to be the case for 790–800 nm. The full optimisation procedure, however, is only performed when the oscillator properties do not allow for reliable amplification. This is typically the case after about two months of day-to-day operation.

3.4.2 Amplification

The stages of the CPA amplifier are shown schematically in Fig. 3.12. Their most important properties are presented below. After the oscillator, the seed beam passes a Faraday isolator that prevents possible light reflections from the amplifier to travel back into the oscillator and cause damage or ML instabilities. The beam is then steered into the pulse stretcher that consists of a single grating and a folding mirror ensuring that the beam hits the grating four times. The throughput efficiency is roughly 50 %. The pulses are stretched to 150 to 200 ps to ensure a sufficiently low intensity during amplification and hence avoiding damage to the amplifier optics. Here, it is crucial that the bandwidth of the input pulse is sufficiently large (>45 nm) to enable stretching to the required pulse length. Additionally clipping or other distortions of the spectrum in the stretcher must be avoided as they may lead to intensity spikes during amplification.

The stretched seed then passes the pulse picking stage, where the repetition rate is reduced to 1 kHz. This is achieved via a Pockels Cell (PC) followed by a polariser. When a high voltage pulse is applied to the PC, the laser pulse polarisation is rotated from horizontal (P) to vertical (S) with respect to the optical table. Only S-polarised pulses pass the subsequent polariser. Triggered at 1 kHz and with the adequate time delay with respect to the seed pulse train, pulses thus enter the amplifier at 1 kHz repetition rate.

The stretched kHz pulse train then enters a first multipass amplifier, which consists of a cryogenically cooled Ti:S crystal pumped by 30 mJ, 532 nm and 100 to 300 ns pulses. The pump is a frequency doubled, Q-Switched, diode pumped Nd:YAG laser (DM30, Photonic Industries). The crystal is placed in a UHV vacuum chamber maintaining a background pressure around 10^{-7} mbar. This avoids depositions on the crystal. Liquid He is used to maintain a crystal temperature of about 75 K during pumping. This ensures a high damage threshold and thermal stability. The induced population inversion in the Ti:S crystal provides the gain for the seed, which passes the crystal 13 times. The last pass is picked up by a folding mirror and steers the beam to the pulse cleaning stage. The timing of the pump pulse with respect to the seed pulse is adjusted to maximise gain and minimise amplification of spontaneous emission (ASE). Under ideal conditions, the seed pulse is typically amplified to 2.5 mJ. This corresponds to a gain factor of about 0.5×10^6. At the same time, ASE is kept below 0.2 mJ.

The pulse cleaning stage is identical to the pulse picking stage, except for a slightly larger time window of the operated PC. This stage removes possible pre- or post-pulses to the main pulse. In this way such unwanted pulse artifacts can be suppressed below 0.1 % with respect to the main pulse. After amplification in the second amplifier, the pulse contrast is typically still below 1 %. The pulse train is now P-polarised again. Both the picking and cleaning PC are optimised to maximise output power and pulse contrast.

The second amplification stage consists of an identical crystal chamber. However, the Ti:S crystal is pumped by a 30 W (DM30, Photonic Industries) and a 40 W (DM40, Photonic Industries) laser with otherwise identical specifications compared to the pump laser on the first stage. The seed pulse only passes the second amplifier twice with a typical output of up to 14 mJ per pulse. This corresponds to a gain factor of 5.6. Due to the exponential increase in gain depletion with increasing seed energy, more passes would not significantly increase the overall gain. After the second pass, the beam is upcollimated approximately by a factor of two to reduce nonlinear optical effects during propagation after pulse compression.

The compressor consists of two gratings and a retro-reflecting mirror, such that the output beam hits each grating twice. For optimal pulse compression, the dispersion induced by the compressor has to precisely compensate the dispersion imposed by the stretcher and transmissive optics in the laser. For this, the grating angle with respect to the input beam has to be approximately matched to the corresponding angle of the stretcher grating. This corrects third order dispersion effects.

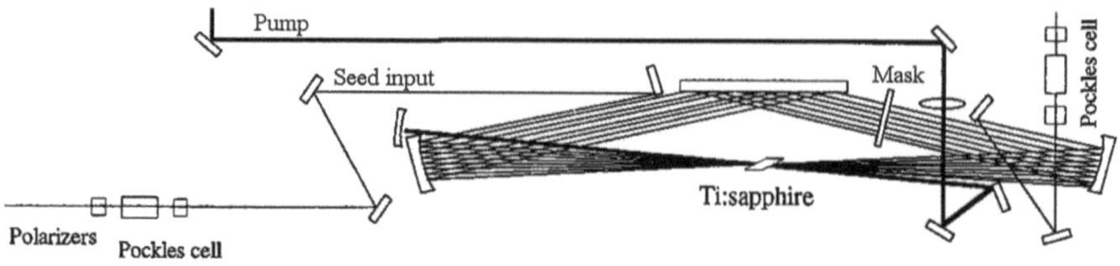

Fig. 3.12 Schematic illustration of the design used for the first amplification stage of the Imperial College CPA system. The displayed part shows the pulse picking and cleaning stages described in the text. Also note that only seven passes through the crystal are shown as opposed to the 13 employed in the presented CPA system. The mask is used for helping with the alignment and spatial mode cleaning. (Reprinted with permission from [4]. Copyright (1998), AIP Publishing LLC)

The distance between the gratings then corrects for second order dispersion effects. Note that the compressor gratings have to be parallel to avoid spatial chirp. The throughput efficiency of the compressor has been determined to be 68 %. Under optimal conditions, this results in a compressed output energy of around 8 mJ per pulse.

3.5 The Imperial College NIR OPA System

The Imperial College near-infrared (NIR) laser is a comercially available system (HE-TOPAS, Light Conversion) that consists of a seed generation and a high energy optical parametric amplification (OPA) stage. It is pumped by the CPA system presented in the previous section and could thus be called a secondary laser source converting 800 nm radiation into a tunable output of 1,200 to 2,600 nm. The conversion process is most efficient and stable for an output centred between 1,300–1,450 nm with up to 1.8 mJ per pulse and a duration of about 30 fs. For stable operation, the Imperial College HE-TOPAS requires 7.6 to 8.0 mJ of 800 nm pump pulses. They are split and delayed inside the system to generate and amplify the NIR output. A pulse length measurement via the previously presented SEA-F-SPIDER apparatus has been performed by Sébastien Weber and is displayed in Fig. 3.13.

3.5.1 Seed Generation

The initial pump beam is split at the input via a beam splitter (BS). Approximately 5 % of the pump energy is used for generating the NIR seed (seed-pump) that is amplified with the remaining pump energy in the high-power amplification stage (HE-pump). The 800 nm seed-pump now carries about 300–600 μJ and is split by another BS such that 15 % of the seed-pump are used for generating the NIR seed. 1–3 μJ of this fraction are now focussed onto a sapphire plate to generate a white-light contiuum

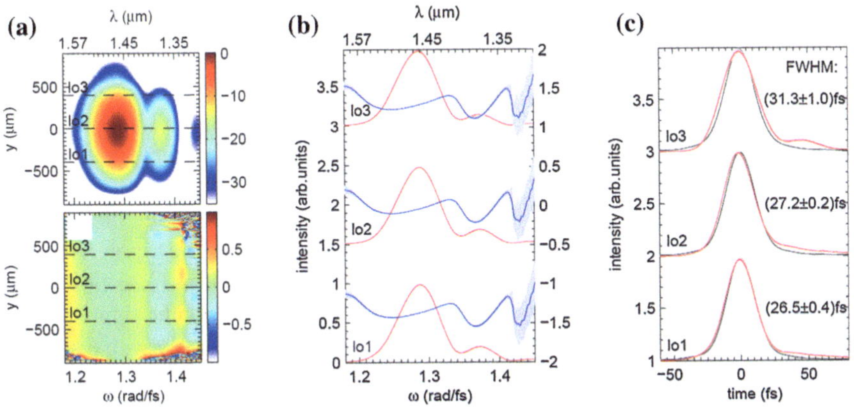

Fig. 3.13 Spatio-temporal reconstruction of a 1.4 μm pulse from the Imperial College NIR OPA system. **a** Spatially resolved spectral amplitude (*top*) and associated relative standard deviation in percent. **b** Spectral lineouts from the positions indicated in (**a**) including the reconstructed phase functions. **c** Reconstructed (*red solid line*) and transform limited pulse (*black solid line*) in the temporal domain corresponding to the spectral lineouts in (**b**). (Reprinted from [28], with permission from Optical Society of America)

(WLC). The tuning capability of the HE-TOPAS comes from the pre-amplification stage that uses the remaining 30–50 μJ for optical parametric amplification of the wavelength selected from the WLC. Here, the WLC and 800 nm pump are focussed non-collinearly into a nonlinear crystal. This simplifies the separation of the amplified signal from the idler and residual pump beam. The crystal angle is tuned via a software controlled motor and selects the amplified signal wavelength. The delay of the pump beam is adjusted via a software controlled, motorised delay stage for maximising the amplification efficiency. The NIR seed is expanded via a lens telescope before further amplification.

Correct WLC generation is crucial for maintaining a good shot-to-shot stability of the HE-TOPAS output. The tradeoff consists of generating a WLC of sufficient intensity for efficient amplification and avoiding instabilities due to too much pump intensity on the sapphire plate. For achieving this, the 800 nm beam generating the WLC can be apertured with an iris before focussing in order to adjust its intensity.

3.5.2 *Amplification*

Amplification of the NIR seed takes place via two additional OPA stages. Both employ a collinear geometry. For the first stage, approximately 85 % of the 800 nm seed-pump (300–500 μJ) are down-collimated via a lens-mirror telescope for increasing the pump intensity. The high power output of the HE-TOPAS is now achieved via the second stage, where the MIR seed is amplified by the 800 nm he-pump (7.0–7.5 mJ). For this seed and pump are telescoped and collimated to match the size of the large nonlinear crystal (≈1 cm^2) for optical parametric amplification. The residual pump is then separated from the amplified NIR beam via a dichroic mirror. For both amplification stages, crystal angle and pump delay can be adjusted via software controlled motors for wavelength tuning and most efficient amplification. In addition to that, the pulse duration of the 800 nm pump is now adjusted for the highest NIR output. This typically requires a slight detuning of the CPA system's compressor such that the pump pulse is shortest at the high energy amplification stage. The amplified NIR output now consists of an S-polarised signal (1,200–1,600 nm) and a P-polarised idler beam (1,600–2,600 nm). As the idler is generated in the two amplification stages, it travels collinearly with the signal. Due to their differing polarisation however, they can be separated efficiently with a suitable wavelength separator.

As the amplifying pump beam in both stages is not focussed, a good mode of the 800 nm pump is crucial for obtaining efficient amplification and a decent NIR output mode. Furthermore, the energy of the HE-pump has to be sufficiently high for driving the last amplification stage into saturation. This is necessary for obtaining a satisfactory shot-to-shot stability of the HE-TOPAS output.

Experience showed however that the stability of the NIR output also sensitively depends on the stability of the 800 nm pump and is typically worse than the latter. This renders experiments involving the Imperial College NIR laser rather time-consuming and quite challenging endeavours. The reason for this is that for achieving a satisfactory NIR beam, the CPA and HE-TOPAS system have to be adjusted very carefully for optimal performance. This typically requires a complete realignment of the CPA amplifiers and HE-TOPAS before starting an experimental campaign. However even in this scenario, shot-to-shot instabilities typically require extended data acquisition times for achieving adequate statistics.

References

1. A. Einstein, Strahlungs-emission und absorption nach der quantentheorie. Deutsche Physikalische Gesellschaft **18**, 318–323 (1916)
2. T.H. Maiman, Stimulated optical radiation in ruby. Nature **187**, 493–494 (1960)
3. P.F. Moulton, Spectroscopic and laser characteristics of ti:al2o3. J. Opt. Soc. Am. B **3**, 125 (1986)
4. S. Backus, C.G. Durfee, M.M. Murnane, H.C. Kapteyn, High power ultrafast lasers. Rev. Sci. Instrum. **69**, 1207–1223 (1998)
5. R. Ell, U. Morgner, F.X. Krtner, J.G. Fujimoto, E.P. Ippen, V. Scheuer, G. Angelow, T. Tschudi, M.J. Lederer, A. Boiko, B. Luther-Davies, Generation of 5-fs pulses and octave-spanning spectra directly from a ti:sapphire laser. Opt. Lett. **26**, 373–375 (2001)
6. T. Brabec, F. Krausz, Intense few-cycle laser fields: frontiers of nonlinear optics. Rev. Mod. Phys. **72**, 545 (2000)
7. F. Frank, *Generation and Application of Ultrashort Laser Pulses in Attosecond Science*. Ph.D. thesis, Imperial College London, (2011)
8. J. Diels, W. Rudolph, *Ultrashort Laser Pulse Phenomena*, 2nd edn. (Academic Press, San Diego, 2006)

9. U. Keller, in *Laser Systems, Part 1*, ed. by G. Herziger, H. Weber, R. Poprawe. 2.1 Ultrafast Solid-State Lasers, vol 11 (Springer, Heidelberg, 2007), pp. 33–167
10. R. Trebino, *Frequency-Resolved Optical Gating: The Measurement of Ultrashort Laser Pulses*, 1st edn. (Springer, New York, 2002)
11. G. New, *Introduction to Nonlinear Optics* (Cambridge University Press, Cambridge, 2011)
12. C. Rullière (ed.), *Femtosecond Laser Pulses: Principles and Experiments* (Springer, Germany, 1998)
13. M. Nisoli, S. Stagira, S. De Silvestri, O. Svelto, S. Sartania, Z. Cheng, M. Lenzner, C. Spielmann, F. Krausz, A novel-high energy pulse compression system: generation of multigigawatt sub-5-fs pulses. Appl. Phys. B Lasers Opt. **65**, 189–196 (1997)
14. C. Hauri, W. Kornelis, F. Helbing, A. Heinrich, A. Couairon, A. Mysyrowicz, J. Biegert, U. Keller, Generation of intense, carrier-envelope phase-locked few-cycle laser pulses through filamentation. Appl. Phys. B Lasers Opt. **79**(6), 673–677 (2004)
15. M. Drescher, M. Hentschel, R. Kienberger, G. Tempea, C. Spielmann, G.A. Reider, P.B. Corkum, F. Krausz, X-ray pulses approaching the attosecond frontier. Science **291**, 1923–1927 (2001)
16. L.E. Hargrove, R.L. Fork, M.A. Pollack, Locking of hene laser modes induced by synchronous intracavity modulation. Appl. Phys. Lett. **5**, 4–5 (1964)
17. A.J. DeMaria, D.A. Stetser, H. Heynau, Self mode-locking of lasers with saturable absorbers. Appl. Phys. Lett. **8**, 174–176 (1966)
18. D.E. Spence, P.N. Kean, W. Sibbett, 60-fsec pulse generation from a self-mode-locked ti:sapphire laser. Opt. Lett. **16**, 42–44 (1991)
19. D. Strickland, G. Mourou, Compression of amplified chirped optical pulses. Opt. Commun. **56**, 219–221 (1985)
20. D. Kane, R. Trebino, Characterization of arbitrary femtosecond pulses using frequency-resolved optical gating. IEEE J. Quantum Electron. **29**, 571–579 (1993)
21. R. Trebino, D.J. Kane, Using phase retrieval to measure the intensity and phase of ultrashort pulses: frequency-resolved optical gating. J. Opt. Soc. Am. A **10**, 1101–1111 (1993)
22. J.A. Armstrong, Measurement of picosecond laser pulse widths. Appl. Phys. Lett. **10**, 16–18 (1967)
23. H.P. Weber, Method for pulsewidth measurement of ultrashort light pulses generated by phase-locked lasers using nonlinear optics. J. Appl. Phys. **38**, 2231–2234 (1967)
24. C. Iaconis, I. Walmsley, Spectral phase interferometry for direct electric-field reconstruction of ultrashort optical pulses. Opt. Lett. **23**, 792–794 (1998)
25. I.A. Walmsley, C. Dorrer, Characterization of ultrashort electromagnetic pulses. Adv. Opt. Photonics **1**, 308–437 (2009)
26. M. Takeda, H. Ina, S. Kobayashi, Fourier-transform method of fringe-pattern analysis for computer-based topography and interferometry. J. Opt. Soc. Am. **72**, 156–160 (1982)
27. T. Witting, F. Frank, C.A. Arrell, W.A. Okell, J.P. Marangos, J.W.G. Tisch, Characterization of high-intensity sub-4-fs laser pulses using spatially encoded spectral shearing interferometry. Opt. Lett. **36**(9), 1680–1682 (2011)
28. T. Witting, S.J. Weber, J.W.G. Tisch, J.P. Marangos, Spatio-temporal characterization of mid-infrared laser pulses with spatially encoded spectral shearing interferometry. Opt. Express **20**, 27974–27980 (2012)

Chapter 4
Experimental Methods and Setup

The experiments presented in this thesis deal with the molecular dynamics following strong field ionisation. This implies that the resulting reaction products—molecular fragments and parent ions—always include charged particles. The experiments were thus based on detecting and identifying ion yields as a function of experimental parameters such as laser intensity and polarisation state. This imposes several requirements for the experimental setup. Firstly, a suitable ion detection system has to be employed. For this thesis this was an ion Time-of-Flight mass spectrometer (TOFMS). Secondly, for studying the interaction of a particular molecule with a strong laser pulse, one has to ensure that the signal from this interaction dominates over possible background signals. This was achieved by conducting the experiments in a high vacuum environment with background pressures below 2×10^{-7} mbar and by delivering the molecular sample in form of a narrow, high density molecular gas beam. The vacuum ensures a low concentration of background molecules such as H_2O in the interaction region, whereas the molecular beam confines the interaction region to a small volume defined by the laser and molecular beam dimensions. Here, a high sample density is present, whilst the gas density outside this region is kept to a minimum due to the collimation of the molecular beam. Both experimental approaches—high vacuum and collimated molecular beam—then informed the specific layout of the utilised vacuum chamber, which consisted of two parts. The initial gas expansion took place in the so-called source chamber which was connected to the interaction chamber via a 1.0 mm diameter hole. This extracted the molecular beam from the gas expansion whilst maintaining the required low vacuum in the interaction chamber. This method is referred to as differential pumping.

The experiments presented in this thesis study the molecular structure dependence of molecular dynamics such as ionisation and fragmentation. However, in a gaseous sample the orientation of the molecules is random such that the interaction is averaged over all orientations. For resolving the molecular structure dependence, the laser-molecule interaction thus has to take place in the molecular frame. To facilitate this, laser induced impulsive molecular alignment was employed, which prepares a rotational wavepacket in the molecular ensemble via a short laser pulse of moderate

M. Oppermann, *Resolving Strong Field Dynamics in Cation States of CO_2 via Optimised Molecular Alignment*, Springer Theses, DOI: 10.1007/978-3-319-05338-7_4,

intensity on the order of 10^{13} Wcm^{-2}. At the appropriate time delay after the pulse the rotational wavepacket rephases, corresponding to an alignment distribution of the molecular ensemble confined to a narrow cone around the laser pulse polarisation. For using this technique in the context of strong field experiments, one requires a so-called pump-probe beamline. Here, a first laser pulse (the pump pulse) prepares the molecular ensemble. In this case the pump pulse induced impulsive alignment. At a variable time delay, a second laser pulse then interacts with the prepared sample. Here, this was a short high intensity pulse on the order of 10^{14} Wcm^{-2}, inducing strong field molecular dynamics. For the optimisation of impulsive alignment, one has to ensure a low rotational temperature of the molecular sample on the order of a few Kelvin. In the experiments presented in this thesis, this was achieved by seeding the molecular sample in an atomic carrier gas, which enhanced the cooling during the initial supersonic gas expansion.

This chapter describes and characterises the above mentioned experimental tools and methods in detail. Firstly, the field of ion detection and imaging is briefly summarised, leading to the description of the employed TOFMS. Secondly, an overview of the experimental setup is presented. As pointed out above, it consisted of three parts; (1) a vacuum system housing the gas sample delivery and TOFMS, (2) a pump-probe beam line for manipulating the laser pulse parameters and (3) an automated data acquisition setup. In addition, an external gas mixing setup was used for providing homogeneous, binary gas mixes. In the remaining sections, the most important experimental methods employed for the main setup are discussed in more detail. This includes gas sample delivery via a pulsed molecular beam and the manipulation of the laser pulse parameters.

4.1 Ion Detection Techniques

With molecular strong field ionisation as the initial process of the studied molecular dynamics, the resulting reaction products—molecular fragments and parent ions—always include charged particles. For this, a TOFMS was used to detect and identify the yields of the reaction products. In this section, TOF spectrometry is placed in the wider context of ion imaging techniques highlighting its advantages and disadvantages in the context of the experiments conducted for this thesis.

4.1.1 Ion Detection

The flight path of charged particles can be manipulated with extremly high accuracy through the use of static electric and magnetic fields. This has enabled the development of sophisticated ion (and electron) detection schemes with the aim to determine the characteristics of charged particles. In the case of ions, these are mass, charge, quantum state and momentum vector. From the point of view of a molecular physicist, this makes it possible to study molecular reactions through detection and analysis of their end products with the aim to infer the undertaken reaction pathway and thus involved quantum states and potential energy landscape [1].

Ionisation

At some stage of the experiment, ionisation of the reaction products is required. In the experiments presented in this thesis, this process is inherent to the strong field effects under study and ions are always created. It should be pointed out however that neutral reaction products would need to be ionised after their creation. One of the most popular laser-based techniques to facilitate this is resonance enhanced multi-photon ionisation (REMPI), where a two-photon transition to an excited state close to the ionisation threshold is followed by an additional, ionising one-photon absorption (see for example [2] and references therein). Tuning the photon energy to the resonant transition then strongly enhances ionisation of particles in the initial lower state of the resonant transition. This selects the detected ions according to this electronic state. As the electronic state of an ion does not alter its flight path in static electric or magnetic fields, such additional spectroscopic techniques are required when determining all characteristics of the product ions.

Detector

The most common contemporary detection schemes involve micro-channel plates (MCPs). They are thin, flat glass discs consisting of numerous glass capillaries reaching from one of the disc's side to the other. Typically, an MCP is about 2 mm thick, with a capillary diameter of about 10 μm, a capillary spacing of ca. 12 μm from centre to centre and a capillary angle of about 8° with respect to the normal of the MCP's face [3]. When a large electric field is applied to the plate, each capillary acts as an electron multiplier or channeltron, due to a dielectric coating on its inner walls. Hence when an ion enters a capillary it triggeres a cascade of electrons, travelling down the capillary. Capillary diameter, spacing and angle are chosen to provide the largest total cross section for an ion hitting the inside of a capillary and for maximising the number of cascades within the capillary. For increased gain, an MCP detector typically consists of two to three plates stacked behind each other with the channels forming a zigzag pattern. This is called a chevron arrangement and simply increases the total number of cascades within the MCP stack. As with the capillary angle, the chevron arrangement ensures that the cascade leaving the previous plate hits the inner wall of a capillary of the next plate as close to the capillary entrance as possible. In this way gain factors of $10^6 - 10^8$ can be achieved and the impact of a single charged particle can be amplified to a measurable current pulse.

The simplest way of detecting the current pulse is via an anode. This provides an excellent resolution of the arrival time of the current pulse with respect to some fixed reference time, as the pulse transmission time through the detector is typically only around 100 ps [3]. Furthermore multiple hits at the same time simply lead to current pulses with higher magnitude. Due to the linear scaling of the detector signal with the number of hits, this does not impose limitations. However, the spatial extent of the MCP allows the retrieval of the impact position of the charged particle on the detector. For this the anode is replaced by a phosphor screen that is imaged with a CCD camera and hence recording the position of the amplified current pulse hitting the phosphor screen. This approach was pioneered by Chandler and Houston [4] and the detector including the phosphor screen is usually referred to as an imaging MCP. The price that is paid when replacing the anode for such an imaging detector is that the information of the ion arrival time is lost due to the phosphor integrating the signal over time. One should note, however, that the voltage drop at the MCP plates upon detection of an ion can still be picked up via an external circuit. This implies that an imaging MCP can still be used to extract time information. Nevertheless, time resolutions comparable to anode based MCPs have not been reported yet.

This problem has been solved through the development of anode types with spatial resolution. There are numerous approaches and they are summarised in the review by Chichinin et al. [3]. The most widely used type today is the delay line anode. Here, two wire pairs (the delay lines) are wound around a quadratic anode plate, such that the wire pairs cross at 90° [5, 6]. The arrival position of the current pulse on this wire mesh can then be inferred from the time it takes the charge to travel down each wire pair. Delay line anodes suffer from very limited multi-hit resolution and the fact that hits at the same position and too close in time may not be distinguished. This typically limits the number of detectable events in a given time frame. These limitations have been reduced however by employing a third wire pair on a hexagonal anode for example.

4.1.2 Time-of-Flight Mass Spectrometry: 1D Imaging

Time-of-Flight (TOF) mass spectrometry was first proposed in 1948 [7] and is based on the fact that the acceleration of a charged particle in an electric field is proportional to the ratio of its charge q and its mass m. When the creation of ions is pulsed, the time it takes them to travel from their source to a detector can thus be correlated with this ratio. Typically, a TOF spectrometer consists of an acceleration region with a homogeneous electric field and a field free drift region towards an MCP detector with an anode.

Principle of operation

TOF spectrometry became an efficient detection technique through the introduction of a second acceleration region by Wiley and McLaren in 1955 [8]. This reduced the constraints on the geometry of the spectrometer and improved its resolution. Single stage acceleration regions are still in use today, because of their simple implementation and other means of improving the spectrometer resolution. In the following the results from Wiley and McLaren's seminal paper are briefly presented, with the single stage acceleration configuration as a special case.

Figure 4.1a displays the geometry of a double stage acceleration region TOF spectrometer. Here the electrodes are constructed as fine conducting meshes, establishing unifrom acceleration fields. The total flight time of an ion created at

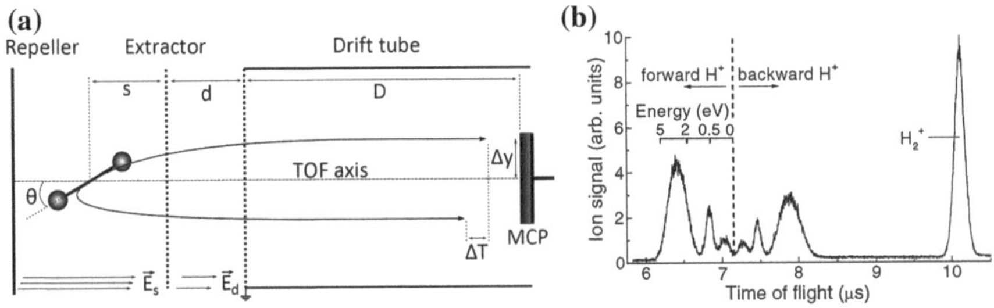

Fig. 4.1 **a** Schematic drawing of a TOF spectrometer with two acceleration regions. The interaction region is at a distance s with respect to the extractor electrode. Note that the dimensions of the displayed diatomic molecule are greatly exaggerated for better visibility. **b** Experimental ion TOF spectrum resulting from the photoionisation of H_2. Note that the H^+ peak is split according to the dissociation channels' kinetic energy release as explained in the text (Part **b** reprinted from [9].

a distance s from the first electrode with initial energy U is the sum of flight times in each spectrometer region:

$$T(U, s) = T_s + T_d + T_D \tag{4.1}$$

where T_s denotes the flight time from the interaction region to the extractor, T_d the time to fly the distance d from extractor to drift tube entrance and T_D the time taken to fly along the length D of the drift tube. As an approximation one may now set $U = 0$, which is justified for ions with an initial energy spread around zero. It then turns out that:

$$\sqrt{\frac{m}{q}} \propto T(0, s) \equiv \sqrt{\frac{m}{q}} = AT(0, s) + B \tag{4.2}$$

In practice, one may use this result to calibrate a TOF spectrometer if two mass peaks in a mass spectrum are known. In this case the two values can be used to determine the free parameters A and B, relating the recorded flight time to the associated mass to charge ratio.

Two effects limit the mass resolution. The initial spatial spread of the created ions around $s = s_0$ and their initial energy distribution around $U = U_0$. In the first case, ions further away from from the first grid ($s > s_0$) will travel a longer distance but acquire more kinetic energy. Setting $U = 0$ and $s = s_0$, the condition $dT/ds = 0$ implies the minimum change of flight time under small changes in s. This space focussing condition then relates the spectrometer parameters as follows [8]:

$$D = 2s_0 k_0^{3/2}\left(1 - \frac{1}{k_0 + \sqrt{k_0}}\frac{d}{s_0}\right) \tag{4.3}$$

where $k_0 = (s_0E_s + dE_d)/(s_0E_s)$. This implies that the static field strengths can be adapted to establish space focussing conditions for any given spectrometer geometry.

In the case of a single stage acceleration region, $d = 0$ and $E_d = 0$, such that Eq. 4.3 reduces to a geometrical constraint with $D = 2s_0$.

The initial energy of the ions spreads their flight time. This can be seen when considering two identical ions, created at the same time and location but with initial momenta of the same magnitude but opposite direction along the TOF axis. In this case one ion initially flies away from the detector, is turned around by the static electric field and arrives later than the ion initially flying towards the detector. This delay or turn-around time is given as [8]:

$$\Delta T = 1.02\frac{2\sqrt{2mU_0}}{qE_s} \tag{4.4}$$

Energy focussing can thus be improved either by reducing U_0 or by increasing E_s. Generally, energy focussing is improved when the initial energy of the ion is kept as small as possible in relation to its total energy at any point along its flight path. Concerning the total TOF resolution, note that by increasing E_s, the flight time is shortened, which leads to a decrease in TOF resolution. Typically a compromise is established by using a long drift tube of about $D \approx 2\,\mathrm{m}$ and choosing acceleration fields that lead to high kinetic energies of the ions around several kV. Energy focussing has further been improved by the use of delayed ion extraction [8, 10] and by the development of a so-called reflectron TOF spectrometer [11].

Momentum sensitivity

As pointed out previously, a TOF spectrometer is sensitive to the ion's initial momentum unless very good energy focussing is employed. Two further perspectives on this issue have to be regarded. First of all, if energy focussing is not established, the spread of the TOF signal can be used to determine the initial kinetic energy spread of the associated ions. In particular, in molecular fragmentation this effect can be used to separate dissociation channels according to their kinetic energy release. As is shown in part (a) of Fig. 4.1 and taking into account Eq. 4.4, the turn-around time depends on the momentum component along the TOF axis. This means that fragmentation channels with high KER lead to ions with a larger separation in TOF. In a typical TOF spectrum this means that the associated ion peak is split in two peaks separated by ΔT. This is displayed in part (b) of Fig. 4.1.

Due to the finite size of the detector, a second effect has to be considered. Any momentum component perpendicular to the TOF axis will lead to a displacement Δy perpendicular to the TOF axis:

$$\Delta y = \sqrt{2U_0/m}\cos(\theta)T(U_0, s_0) \tag{4.5}$$

This implies that ions may miss the detector depending on their initial momentum component perpendicular to the TOF axis. This effect can be minimised by either using a larger detector or by decreasing the TOF of the ions through shorter drift distances or higher acceleration fields. For a given set of the above mentioned parameters and a given dissociation channel, there is thus a maximum ejection angle

θ_{max}, such that fragment ions ejected at $\theta > \theta_{max}$ miss the detector. In the case of electrons, this effect has also been circumvented by the use of magnetic fields as will be presented later.

Intensity selective scanning

In a TOF spectrometer, the entire interaction region where ions are created is imaged. In the case of ionisation in a laser focus, this implies that a range of intensities participate in the ionisation process. This is due to the nonlinear 3-dimensional intensity distribution within the focus. The resulting focal volume averaging is strongest in the propagation direction of the laser beam, which offers a convenient way to limiting this source of uncertainty in laser intensity. Due to the spatial separation of different intensity regions within the focus, one may insert a narrow slit in front of the TOF drift region and thereby limit the detection to a certain intensity slice of the focus [12, 13]. This also offers a way of varying the effective intensity by translating the focus along its propagation direction in front of the slit. This technique has been coined *intensity selective scanning* [14] and has been applied together with numerical deconvolution procedures for high precision ionisation [15] and dissociation rate measurements (see for example [16]).

Covariance mapping

Covariance mapping is a statistical method that allows for the correlation of different ions within a TOF spectrum. In the context of molecular dynamics research it was first applied to CO [17] for distinguishing between different ionisation channels. For instance, an O^+ ion may result from either of the two dissociation channels $[CO^{++}] \rightarrow C^+ + O^+$ and $[CO^+] \rightarrow C + O^+$. In the first case, however, both a C^+ and an O^+ ion appear in the TOF spectrum. Such correlated events are extracted via the covariance mapping technique. The branching ratio between the above mentioned channels can then be obtained by comparing the number of correlated O^+ ions to the total number of recorded O^+.

In order to extract covariances between events, high statistics are required. Typically on the order of 10^5 ionisation events have to be acquired in the form of individual TOF spectra per laser shot. Furthermore, the detection efficiency has to be as high as possible, because undetected ions would disturb the extraction of covariances. The main advantage of this technique is that there is no upper limit on the number of detected events per TOF spectrum. This is because covariance mapping is a purely statistical data analysis method and only requires that all ions from a single laser shot are recorded. Covariance mapping has been extended from diatomics to triatomic molecules [18, 19] and has also been applied to photoelectron spectroscopy [20]. Recently the technique has also been demonstrated for angularly resolved photo-ion spectra obtained with a velocity map imaging spectrometer [21].

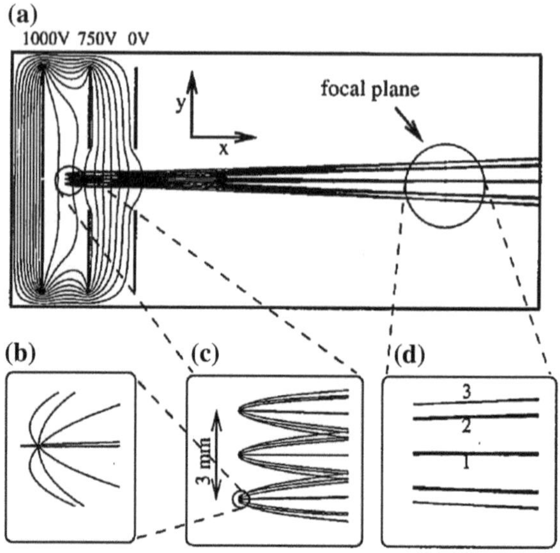

Fig. 4.2 **a** Electrostatic lenses and resulting equipotential lines in a typical VMI spectrometer. The velocity focussing of such an arrangement is shown by numerical simulation of ion trajectories in *insets* **b–d** (Reprinted with permission from [22]. Copyright (1997), AIP Publishing LLC)

4.1.3 Velocity Map Imaging: 2D Imaging

2-dimensional ion imaging was first realised by Chandler and Houston, by replacing the spatially integrating anode by a phosphor screen within the MCP detector [4]. This enabled recording information about the angular distribution of the recorded ions. Eppink and Parker then modified the acceleration region in the spectrometer by replacing the grid electrodes by open ring electrodes. The resulting electrostatic lenses still enable space focussing conditions like in the Wiley-McLaren settings. However the curved equipotential surfaces now also provide space focussing in y-direction (perpendicular to the TOF axis). This leads to ions with the same momentum vector being projected onto the same position at the detector. The technique has therefore been coined Velocity Map Imaging (VMI) and is shown in Fig. 4.2.

Without further modifications of the setup, a VMI records a 2-dimensional projection of the 3-dimensional ion cloud created in the interaction region. The 3-dimensional distribution can be reconstructed, however, if cylindrical symmetry of the ion distribution around the TOF axis can be assumed [23–25]. The main advantage of VMI is that it enables the measurement of angular distributions of ions or electrons within a single laser shot. However, TOF information is lost and 3D distributions can usually not be recovered if the studied interaction does not obey cylindrical symmetry considerations. In this latter case the VMI's velocity and hence energy resolution is substantially reduced due to the superposition of ions with different momentum vectors at the detector. This problem has been avoided by using tomographic imaging procedures (see for example [26, 27]).

4.1.4 Further Techniques: 3D Imaging and Beyond

Several techniques have been developed in order to obtain complete information about particle reactions. The most important ones are briefly presented in this section.

3D particle imaging

In order to obtain the full information in an ion imaging experiment, the spatial distribution of the ions has to be measured together with their TOF. There are two strategies for achieving this. Firstly, one may equip an anode based MCP detector with spatial resolution, for instance a delay line anode [28]. The second possibility is to add time resolution to an imaging MCP detector, which requires a time-sensitive read-out system of the phosphor screen. This has been achieved by imaging the phosphor screen with two independent and differently timed CCD cameras [29] or by using a single fast, double exposure CCD camera [30]. Even though more sensitive to multi-hit limitations, the first strategy has become the more popular choice due to much higher timing resolution.

Coincidence detection

In order to study reaction pathways following molecular ionisation, PhotoElectron-PhotoIon COincidence (PEPICO) spectroscopy was developed in the 1970s [31, 32]. Briefly, a coincidence spectrometer consists of an ion and an electron TOF spectrometer facing each other and sharing a common interaction region. If only one molecule is ionised at a given time, the detected product ion(s) and photo-electron(s) are correlated. By employing single-photon ionisation, the kinetic energy of the recorded electron is given as $E_e = \hbar\omega - I_p$, where ω is the photon's angular frequency and I_p the probed ionisation potential of the molecule. The parent ion thus carries an internal excitation energy of $E^*_{ion} = I_p - I_0$, where I_0 is the lowest ionisation potential of the ionised molecule. By resolving the recorded photoelectron spectra in terms of their kinetic energy, one can thus extract the product ion's internal energy and reaction pathway.

In practice, sources of background ions and electrons need to be kept as low as possible to avoid false coincidence events [33]. Furthermore the collection efficiency of the spectrometers must be as high as possible so true coincidences are not missed. The main disadvantage of this technique is the low number of events for ensuring that only one molecule is ionised at a given time. However, if some kind of energy analyser is used for recording the kinetic energy of the photoelectrons, instead of an electron TOF, continuous ionisation sources can be used, as the recorded electron provides the time-stamp for the ion TOF spectrometer. This may significantly enhance data acquisition rates. Coincidence studies have also been angularly resolved by replacing the TOF spectrometers by 3D imaging spectrometers (for a review see [33] and references therein).

COLTRIMS

Cold target recoil ion momentum spectroscopy (COLTRIMS) was developed in the 1990s [35, 36] with the aim of a kinematically complete description of atomic or

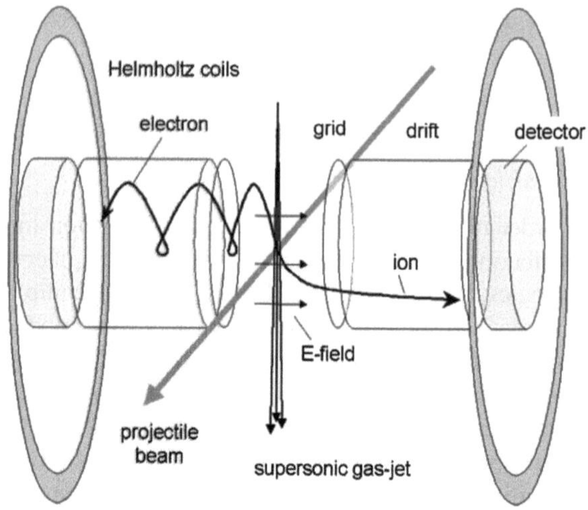

Fig. 4.3 Schematic drawing of a reaction microscope as explained in the text. The projectile beam refers to the collision partners inducing the atomic or molecular reaction. The beam typically consists of electrons or photons (Reprinted from [34], with permission from Springer Science+Business Media)

molecular reactions (for a review see [34]). The corresponding spectrometer is often termed a *reaction microscope* and is essentially a coincidence spectrometer equipped with 3D imaging detectors, a complete acceptance angle of 4π steradians and a sample delivery via a supersonically cooled molecular beam. A schematic drawing of a reaction microscope is displayed in Fig. 4.3. The electron and ion spectrometers face each other and share a common interaction region and typically utilise position sensitive delay line anodes. This allows recording the momentum vectors of all reaction products. High collection efficiency of the electrons is ensured by employing a weak uniform magnetic field along the central spectrometer axis, provided by Helmholtz coils. The magnetic field forces the electrons onto helical trajectories and thus effectively confines them in a cylindrical space around the spectrometer axis. This avoids electrons missing the detector. By utilising supersonic molecular beams (see Sect. 4.4), the atomic or molecular targets are cooled down to the few milliKelvin level. This greatly reduces the thermal energy distribution of the target particles and thus their random momentum distribution. This enables an ion momentum resolution at the 1 μeV level and 1 meV for low energy electrons (< 5 eV) [34]. For coincidence detection, electrons and ions can be directly correlated via momentum conservation. In the case of molecular dissociation, the detection of all fragments therefore also allows the reconstruction of the molecular orientation within the reaction microscope.

COLTRIMS is currently regarded as the state-of-the-art experimental technique for studying atomic or molecular reactions, as it enables a complete description of the reaction pathway. However, being a coincidence detection technique, it still suffers from very low data acquisition rates. Furthermore the delivered raw data is extremely complex due to the full description of each reaction product. This renders the data analysis an extremely time consuming challenge.

4.2 The Imperial College TOFMS

The experiments presented in this thesis require extensive scans of many different laser parameters. This means that fast data acquisition rates are desired. Furthermore the formation rate of different ion species have to be monitored at the same time. This includes recording single and double ionisation rates simultaneously, monitoring different fragment ion species formed in the same laser shot and monitoring atomic reference peaks at the same time. These requirements can be fulfilled with a TOF mass spectrometer, but not with a standard VMI arrangement. Coincidence techniques would provide more information, yet data acquisition rates would be severely reduced. These are the main reasons for employing a TOFMS for the experiments presented in this thesis. The used spectrometer is presented in this section.

4.2.1 Spectrometer Description

The TOFMS used for the experiments in this thesis is shown in Fig. 4.4. It had been constructed as an electron VMI spectrometer and thus had an open electrode configuration and a μ-metal shielding to avoid the influence of the earth's magnetic field onto the electron trajectories. The VMI's position sensitive detector, however, was replaced by an MCP detector with a spatially integrating anode. The detector was a chevron two-stack MCP with a 15 mm diameter active area (F4655-12FX-114, Hamamatsu). When used at maximum allowed voltage settings it provided a gain factor on the order of 10^6. The distance between the repeller and extractor electrode was 15 mm. The same distance was chosen for the distance between the extractor and the drift tube entrance. The openings of the repeller and extractor electrodes had a diameter of 20 and 24 mm, respectively. With the interaction region located at the center between repeller and extractor, the total distance to the detector was 136.5 mm.

Marco Siano performed ion trajectory simulations using the commercially available software SIMION to determine the space focussing settings of the spectrometer. For this, N^+ ions were propagated from the interaction volume to the detector as a function of the voltage ratio $R = V_E/V_R$ between the extractor and repeller. The cylindrical interaction volume had a diameter of 100 μm and a length of 2 mm. The results of this calculation are displayed in Fig. 4.5 where the standard deviation σ_{TOF} of the calculated TOF of the launched ions is plotted as a function of R. Space focussing was achieved for $R = 0.71$. Space focussing conditions can be checked in situ by observing the width of atomic ion TOF peaks and the peak splitting of molecular fragment ions. Keeping the repeller voltage fixed, the extractor voltage is varied for obtaining the sharpest atomic ion TOF peaks. This corresponds to optimal space focussing. Furthermore, space focussing conditions also imply that fragment ion peaks are split symmetrically in time, which provides another diagnostic for experimentally determining the appropriate electrode voltages. On a day to day basis the

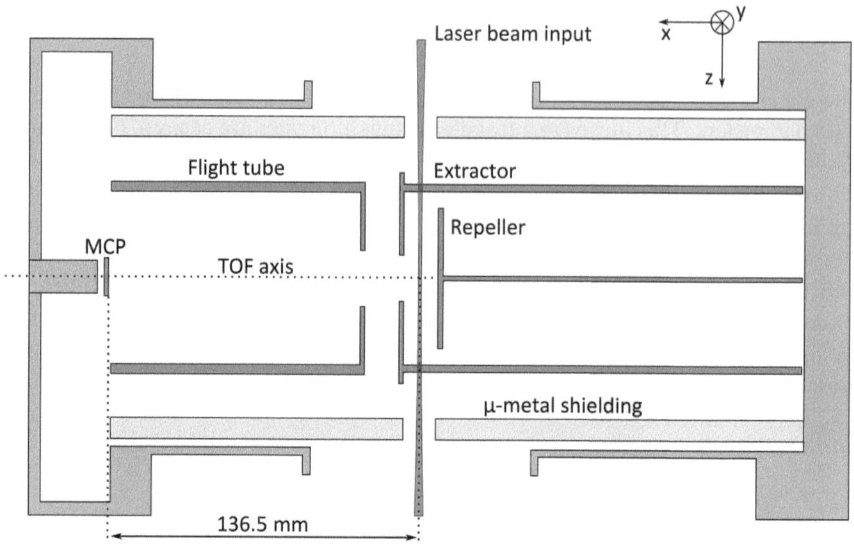

Fig. 4.4 Schematic drawing of the Imperial College TOFMS

Fig. 4.5 Simulated standard deviation of arrival TOF of N^+ ions as a function of the electrode voltages in the Imperial College TOFMS

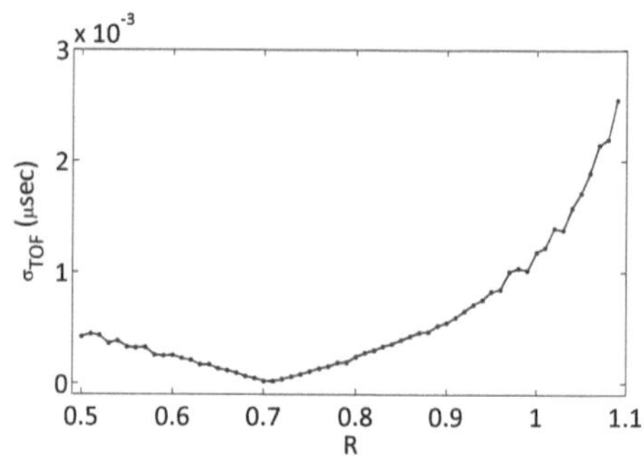

latter approach was used, monitoring the peak splitting of H^+ from background H_2 and H_2O molecules. In this way, the calculated ratio for R could be confirmed.

For all experiments in this thesis, the electrode voltages were set to $V_R = 1.2\,\text{kV}$ and $V_E = 0.85\,\text{kV}$. The voltages were chosen to be sufficiently low for achieving a peak splitting that is large enough to separate relevant dissociation channels. Lower voltages also decrease the acceptance angle of the detector with respect to the ejection angle of molecular fragment ions with respect to the TOF axis. This effect is central to the detection of molecular alignment as is discussed in the following chapter. The MCP was operated with $-2.4\,\text{kV}$ on the front plate and the back plate grounded. The anode was operated at $+0.4\,\text{kV}$. The TOF signal was read out via a digital storage oscillocope (DPO7254, Tektronix).

4.2.2 *Acquisition of TOF Spectra*

For optimal data acquisition, one has to ensure that in a TOF spectrum resulting from a single laser pulse, single ion hits are recorded while the highest TOF peak is not saturated. Assuming that the recorded ion signal is below the saturation threshold of the MCP detector, signal saturation is avoided by simply increasing the voltage scale on the oscilloscope. One then has to increase the MCP gain, such that a single ion hit leads to a TOF peak that is sufficiently high above the noise floor. Here, a signal to noise ratio of 10:1 is ideal, yet in the presented experiments this was limited to a ratio of 4:1. For the utilised detector, single ion hits had a width of approximately 1 ns. For accurate sampling, a TOF resolution of 200 ps per sampling point was sufficient. All acquired TOF spectra consisted of 10,000 sampling points and thus a time window of 2000 ns.

If the above conditions are satisfied, the signal to noise ratio of a TOF spectrum averaged over several single shot spectra is only limited by the number of averaged spectra. This is illustrated in Figs. 4.6 and 4.7, where 5000 single shot spectra were averaged, such that a dynamic range of more than 10^3 was achieved. The spectra were obtained from ionising a gas mix consisting of 10 % O_2 seeded in Ar.

Apart from acquiring individual single shot TOF spectra, one could also average the TOF spectra on the oscilloscope and acquire only the averaged trace. In this case, however, the signal to noise ratio was limited by the dynamic range provided by the oscilloscope. Therefore this acquisition mode was only chosen when the ion count rates per laser shot were suffciently high and large dynamic ranges were not required. This was the case for the impulsive alignment measurements presented in Chap. 5.

4.3 Experimental Setup

In this section an overview of the experimental setup is provided. All experimental results presented in this thesis were produced via this setup.

4.3.1 *Vacuum System*

The vacuum system employed in the experiment is displayed in Fig. 4.8 and consisted of two separate chambers connected via a 1.0 mm diameter skimmer (Model 2, Beam Dynamics Inc). A photograph and schematic drawing of the utilised skimmer is displayed in inset (c) in the same figure. The two chambers could be isolated from each other via a hydraulic gate valve (ZA-E-GV-4000M-P, MDC). In the following, the upper part is referred to as the source chamber and the lower part as the interaction chamber. The pressure in each chamber was monitored via full range gauges that consisted of a Pirani and a cold cathode gauge (PKR 251, Pfeiffer).

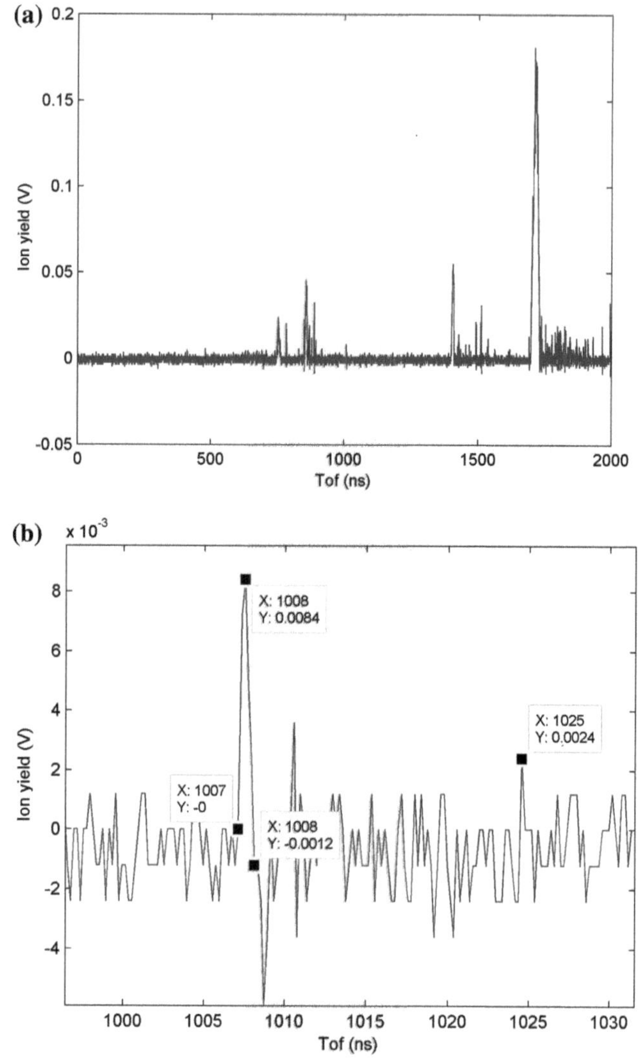

Fig. 4.6 Illustration of ideal data acquisition conditions for the TOFMS used in this thesis. 10 % O_2 in Ar was ionised with 800 nm, 30 fs pulses at an intensity of $(5 \pm 2) \times 10^{14}$ Wcm^{-2} with the *linear* laser polarisation *parallel* to the TOF axis. **a** displays a TOF spectrum from a single laser shot, showing that the maximum peak is not saturated. **b** shows a zoom-in onto the single shot spectrum and shows that single ion hits are resolved above the noise floor with a width of about 1 ns

Source chamber

The source chamber was a 100 mm diameter, 270 mm long steel tube that housed the valve for the gas sample delivery. It was evacuated by two turbo-molecular pumps providing a total pumping speed of about 740 l/s (N_2) (Turbo 1: TPU 261 PC, Pfeiffer

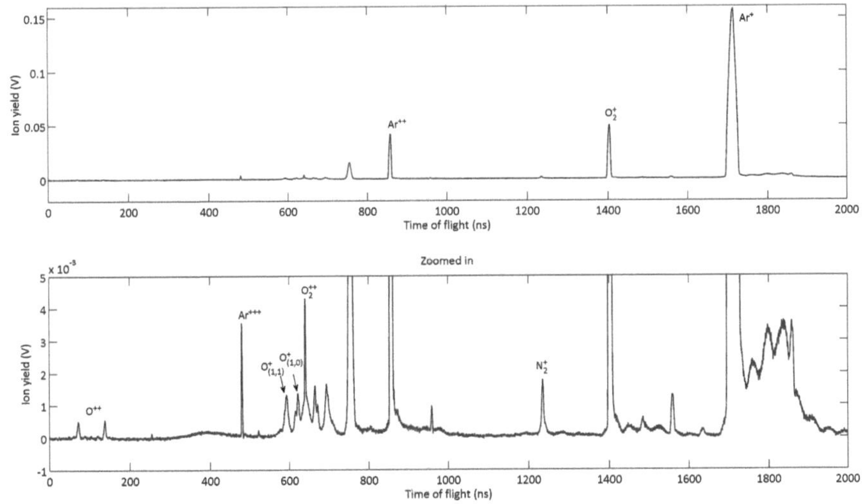

Fig. 4.7 Illustration of ideal data acquisition conditions for the TOFMS used in this thesis. 10 % O_2 in Ar was ionised with 800 nm, 30 fs pulses at an intensity of $(5 \pm 2) \times 10^{14}$ Wcm^{-2} with the *linear* laser polarisation *parallel* to the TOF axis. The figure displays a TOF spectrum averaged over 5000 single shot spectra. The zoom-in illustrates the achieved dynamic range

and Turbo 2: V551, Varian). They were backed by a rotary vane pump providing a flow of approximately 330 l/min (Duo20 MC, Pfeiffer). As this chamber housed the initial gas expansion, efficient pumping was crucial. This can be achieved by increasing the total pumping cross section through the use of turbo pumps with large pumping ports and by increasing the total number of them. For this the largest available turbo (Turbo 1) was mounted on this chamber part. Without the gas valve operating, base pressures at $P_s < 1 \times 10^{-7}$ mbar were routinely achieved.

The gas valve employed for the gas sample delivery consisted of a homebuilt stainless steel reservoir housing a piezo electric disk translator (P-286, PI). A schematic drawing is displayed in inset (a) of Fig. 4.8. The valve was mounted on a manual x-y-z-stage at the top port of the source chamber (not displayed). This allowed for positioning of the valve inside the chamber. The reservoir was filled with a gas sample at a given backing pressure P_b via an external gas line. When no voltage was applied to the piezo disk, the valve was sealed via a stainless steel, rubber seal poppet. When a positive voltage was applied to the piezo disk, the poppet moved up and gas could flow from the valve's reservoir through a nozzle into the source chamber. Here, the poppet's opening distance was proportional to the applied voltage. The valve was operated via the amplified output (HPZT-Power-Amplifier, PI) of a digital delay generator (DG 535, Stanford Research Systems). By externally triggering the delay generator via the laser system electronics, the gas delivery was synchronised with the pulsed laser emission. The maximum backing pressure applicable to the piezo disk was limited to 4 bar. The minimum valve opening time was achieved when a 120 μs long voltage pulse was applied to the piezo disk. This corresponded to an

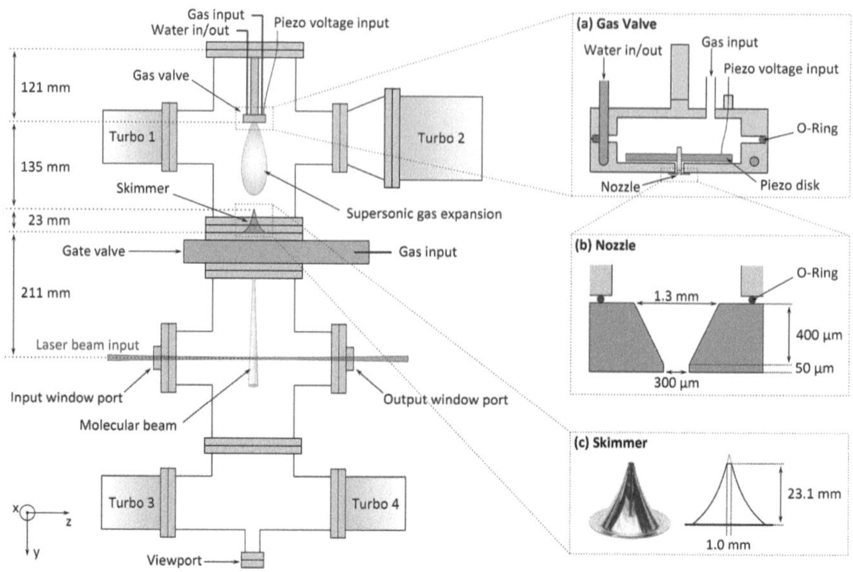

Fig. 4.8 Schematic drawing of the vacuum system employed for the experiments in this thesis. Note that the central axis of the detector is along the x-axis direction, through the interaction region where the laser crosses the molecular beam. The insets (**a**–**c**) illustrate the gas valve, nozzle and skimmer in more detail

effective opening time of 30 μs according to the manufacturer. The maximum possible repetition rate was 1 kHz, which required water cooling of the valve's body. However for the experiments presented in this thesis, the repetition rate was reduced to 47 Hz, which allowed for stable operation without the use of external water cooling.

The nozzle used for the valve was homebuilt and provided a converging shape in the direction of the flow with an exit diameter of 300 μm. Its dimensions are displayed in inset (b) in Fig. 4.8. When the valve was opened, the gas sample expanded into the source chamber. The skimmer then selected the central part of the expansion, resulting in a thin molecular beam travelling into the interaction chamber. The details of the gas expansion dynamics are discussed in in Sect. 4.4.

Interaction chamber

The interaction chamber consisted of a six-way steel cross-piece. It was composed of a 152 mm diameter, 302 mm long tube in x-direction and 100 mm diameter, 270 mm long tubes in z- and y-direction. The chamber housed the Time-of-Flight mass spectrometer with the laser and molecular beam crossing at 90° at its centre. It was evacuated via two turbo-molecular pumps providing a total pumping speed of about 560 l/s (N_2) (Turbo 3: Turbovac 151, Oerlikon Leybold Vacuum and Turbo 4: Turbovac 351, Oerlikon Leybold Vacuum). They were connected to the chamber at its bottom port via a 100 mm diameter cross-piece. The pumps were backed by a rotary vane pump providing a flow of 10 l/min (RV5F, Edwards). Without the gas valve operating, base pressures at $P_i < 5 \times 10^{-8}$ mbar were routinely achieved.

The laser beam entered the interaction chamber via the input window. For every experiment, an 800 nm pump pulse was employed in conjunction with a probe pulse. When the 800 nm pulses from the CPA system were used as a probe, a 1 mm thick, broadband (620–980 nm) anti-reflection (AR) coated fused silica (FS) window (Layertec) was used. For the use of 1350 nm probe pulses from the NIR OPA, a 3 mm thick, uncoated fused silica window (Eksma Optics) was used. Here, an AR coating was not provided, as broadband coatings covering 800 nm and 1350 nm were not available.

A windowed viewport at the bottom of the interaction chamber was installed for aligning the gas valve onto the skimmer. For this, a thin-wired cross-hair was placed onto the view port. A HeNe laser beam was then aligned onto the cross-hair such that it passed through the skimmer. Via a viewport on the source chamber (not displayed), the position of the beam on the valve's body was visible. The valve was then translated for placing the nozzle onto the laser beam.

Differential pumping

For the experiments presented in this thesis, the density of the gas sample at the nozzle exit is desired to be as high as possible, as this leads to the highest rotational cooling (for details see Sect. 4.4). However, the density of molecules in the chamber is limited by: (1) the performance of the TOFMS and (2) the performance of the molecular beam. Firstly, if the density of molecules in the interaction region is too high, the incoming laser pulse creates so many ions that they repel each other due to their Coulomb field. This so-called *space-charge effect* distorts the ion spectra recorded via the TOFMS. Secondly, a molecular beam is only formed if the molecular density in the beam is significantly higher than the surrounding background gas.

The first limitation was tackled by employing differential pumping. As the source and interaction chambers were connected via a 1 mm diameter hole in form of a skimmer, the flow between the two chamber parts was limited. For this, the source chamber could be operated at much higher background pressure than the interaction chamber. When operating the gas valve, the pressure in the interaction chamber was kept at $P_i < 2 \times 10^{-7}$ for avoiding significant space charge effects. This allowed for a maximum pressure in the source chamber at $P_s < 8 \times 10^{-5}$. With the gas valve operating at 47 Hz, minimum opening time and a backing pressure of approximately $P_b \approx 2$ bar, the opening voltage of the valve was tuned for not exceeding these pressures.

Ensuring a low background pressure in the interaction region also provides the best conditions for obtaining a molecular beam. This was checked via additional measurements presented in Sect. 4.4.

4.3.2 Pump-Probe Beam Line

The optical setup used in this thesis was a Mach-Zehnder type interferometer for conducting pump-probe experiments. Its general layout was the same for experiments employing 800 and 1350 nm probe pulses. In the following, the setup is described for the 800 nm case. Modifications for the use of 1350 nm pulses are presented afterwards.

Setup for 800 nm probe pulses

A photograph of the setup including the beam paths is displayed in Fig. 4.9. The input beam consisted of 800 nm, horizontally (P-) polarised pulses at approximately 3.5 mJ, obtained directly from the CPA system. For this, the second amplification stage of the CPA laser was run with only one of the two available pump lasers. The CPA output was expanded to a beam diameter of approximately 25 mm via a reflective Galilean telescope upon exiting the laser box (not displayed). It consisted of a convex, 2 m radius of curvature (ROC) followed by a concave, 4 m ROC silver mirror (Eksma Optics). Doubling the beam size was necessary for propagating the high energy beam at a quarter of the original intensity and thus avoiding nonlinear propagation effects such as self-focussing and self-phase modulation.

The pulses were split via a 45° angle of incidence (AOI) 50/50 beam splitter (BS) into a probe and pump arm. The BS was 1 mm thick, broadband coated (750–850 nm) and made of FS (Layertec). The pump pulse was used to induce molecular alignment (see Chap. 5) and is therefore referred to as the alignment pulse.

The probe pulse passed a translation stage (S) (Owis) that was used for temporally delaying the probe with respect to the alignment pulse. It was further propagated through an attenuator consisting of an achromatic (650–1100 nm) half-wave plate (HWP1) (Altechna) and a broadband (750–850 nm) thin film polariser (P) (Altechna). The attenuator was used for manipulating the probe pulse energy and was software-controlled via the motorised half-wave plate (URS100BCC, Newport). The probe pulse then passed an achromatic (650–1100 nm) quarter-wave plate (QWP) (Altechna) for manipulating its ellipticity. The plate was software-controlled via a motorised rotation stage (PRM1Z8, Thorlabs). Through the use of magnetic post holders, the quarter-wave plate could easily be exchanged for an achromatic (600–1100 nm) half-wave plate (HWP2) mounted in an identical rotation stage. In the photograph QWP is placed in the probe arm and HWP2 stored next to it. The probe pulse was then transmitted through a 0° AOI 20/80 beam combiner (BC). The BC was 1 mm thick, broadband coated (750–850 nm) and made of FS (Layertec). The pulses were then focussed into the interaction chamber by a 30 cm focal length, plano-convex lens made of uncoated BK7 (Newport). The lens was mounted on an x-y-z-stage allowing for the fine tuning of the lens position. The optimal lens position corresponded to the focus hitting the centre of the gas jet and being placed on the central spectrometer axis. As was mentioned previously, a 1 mm thick, broadband AR coated FS window (Layertec) was used at the chamber input.

Note that there were no reflective optics between the probe polarisation optics and the interaction region in the vacuum chamber. This avoided distortions of the

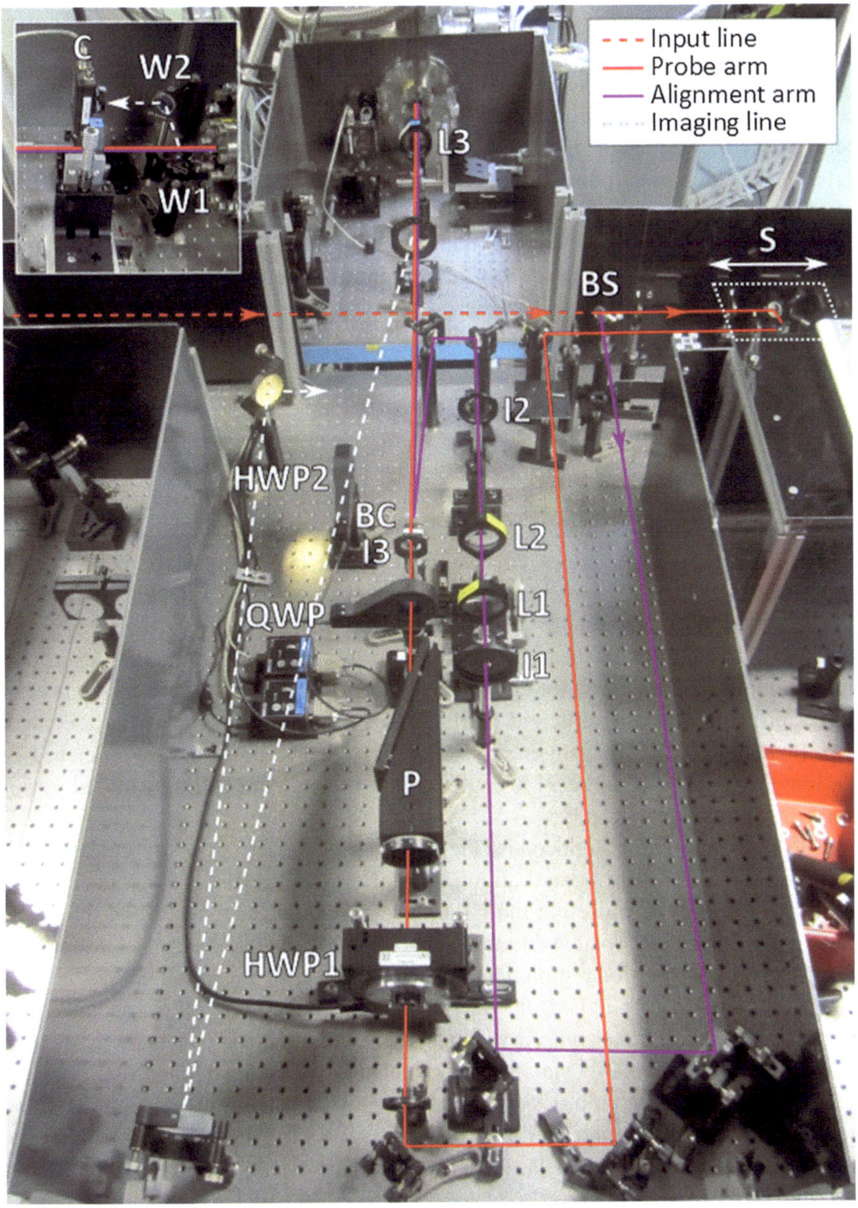

Fig. 4.9 Photograph of the optical setup employed for the experiments presented in this thesis when using 800 nm probe pulses. The *inset* displays the setup for sampling the laser focus via a CCD camera. For details and meaning of the labels see text

polarisation state due to misaligned optics. For the same reason, only one wave plate (either QWP or HWP2) was used in the probe arm at all times.

The alignment pulse was propagated through an iris (I1) used for attenuating the beam in energy and width. Both effects resulted in a reduced intensity when focussed, because the focal spot size increases with decreasing beam width (see Sect. 4.5.1 for details). The alignment pulse then passed a Galilean telescope consisting of a 20 cm focal length, plano-convex (L1) and a 10 cm plano-concave (L2) lens, both made of BK7 (Newport). L2 was placed on a linear translation stage for adjusting its distance from L1. The telescope was used to further reduce the beam size and controlling the divergence of the alignment beam. For measurements requiring the manipulation of the linear alignment pulse polarisation, HWP2 was placed in the beam path after the telescope. The alignment pulse was then propageted to the BC, where 20 % of its energy was combined with the probe pulse.

Typical pulse energies after the BC were measured to be 600–750 μJ for the probe and 110–150 μJ for the alignment pulse. These energies were significantly lower than expected given the input energy and the splitting and combination ratios. The additional loss was mainly due to the need for expanding the input beam to approximately 25 mm diameter. This was necessary to avoid propagation effects. The large beam diameter, however, lead to significant clipping of the beam on the BS as the first 1 inch diameter optic at 45° AOI. For the experiments conducted with this setup, however, this was not an issue as the obtained probe and alignment energies exceeded the minimum requirements (for details see the following experimental chapters) and the resulting foci were of sufficient quality. Also note that the lens was not AR coated, such that some of the incident energy was reflected. The transmission efficiency of the lens was measured to be 91 % by recording the incident and transmitted power of the 800 nm beam.

Due to the dispersion induced by the air and transmissive optics in the beamline the pulse duration of probe and alignment pulse are stretched. This was compensated by chirping the pulse via the CPA system's compressor, such that the probe pulse had the shortest possible duration. This typically resulted in a 30 fs probe pulse duration at the interaction region. Under these conditions, the alignment pulse duration was approximately 70 fs. Both were measured using the SPIDER apparatus (see Sect. 3.3.3). The alignment pulse was longer due to higher dispersion in the associated beam line. This was mainly caused by the transmissive Galilean telescope.

For convenience, beam blocks were installed in the beam line that could be tilted in and out of the beam path. In this way the probe beam could be blocked after the first mirror following the translations stage. The alignment beam could be blocked after the BS.

Setup for 1350 nm probe pulses

When using 1350 nm probe with 800 nm alignment pulses, the optical setup had to be modified to accomodate the different spectral bandwidth and beam splitting and combination requirements. Firstly, the CPA system was operated with both pump lasers on the second amplification stage. This resulted in an ouput energy of approximately 8.5 mJ. The 800 nm pulses were split via a 0° AOI beam sampler with approximately

95 % reflectivity. In this way approximately 7.8 mJ were used for pumping the NIR OPA, which resulted in an output energy of 1.6–1.7 mJ at 1350 nm. The remaining 0.6 mJ at 800 nm from the CPA system were used for the alignment beam.

The probe pulses were coupled into the beam line via a mirror before HWP1. Different polarisation optics were then used to account for the 1350 nm centre wavelength of the probe pulse. HWP1 was an achromatic (1100–2000 nm) half-wave plate (Thorlabs) and was used with a broadband (1300–1400 nm) thin film polariser (Altechna). QWP was an achromatic (1100–2000 nm) quarter-wave plate (Thorlabs), while HWP2 was not used in the probe arm. The probe then passed through a 0° AOI dichroic mirror used as a beam combiner (BC). The BC was a 5 mm thick UV fused silica (UVFS) substrate with an AR coating at 1100–1600 nm and a broadband high-relection (HR) coating at 800 nm (Altechna). The probe pulse then passed a 25 cm focal length, plano-convex lens made of uncoated BK7 (Newport). A shorter focal length lens was used to compensate for the longer wavelength and smaller beam diameter of the 1350 nm probe compared to the 800 nm one and thus achieve similar focal spots in both cases. As was mentioned previously, the focussed beam entered the interaction chamber via a 3 mm thick, uncoated fused silica window (Eksma Optics).

Upon exiting the CPA box, the 800 nm alignment pulse was propagated through the reflective Galilean telescope and thus expanded to a beam diameter of approximately 25 mm. It then travelled along the input beam line through translation stage (S), with the BS removed. Note that with this arrangement, the alignment pulse was delayed with respected to the probe pulse. An additional mirror was inserted after S to couple the beam into the alignment arm. Here, the optics were not changed, as the alignment pulse was kept at 800 nm. As was mentioned previously, HWP2 could be placed into the alignment beam line after L2. This was done for all experiments involving 1350 nm probe pulses. The alignment beam was then propagated to the BC where it was overlapped with the probe beam.

Typical pulse energies after the BC and before the lens L3 were measured to be 650–750 μJ for the probe and 250–300 μJ for the alignment pulse. Again, these energies were significantly lower than the respective input beam energies of 1.6–1.7 mJ and 600 μJ. In the case of the probe, most of the energy was lost due to clipping at the half-inch diameter wave plates. Furthermore the transmission of the polariser was specified at approximately 85 % and thus further reduced the available probe energy. The alignment pulse energy was significantly reduced due to its large beam diameter of approximately 25 mm. This lead to significant clipping on the utilised 45° AOI 1-inch diameter mirrors. In both cases, however, the final pulse energies were sufficient for conducting the experiments and thus modifying the setup for increased energy transmission was not necessary.

Due to using an uncoated focussing lens and an uncoated interaction chamber entrance window, the pulse energy transmitted into the chamber was significantly reduced. The final transmission was determined by measuring the pulse energy before and after an identical lens and laser window. This resulted in 85 % transmission for 1350 nm and 84 % for 800 nm radiation.

Typical pulse durations of the 1350 nm probe were measured to be 30 fs via the NIR SEA-F-SPIDER apparatus (see Sect. 3.3.4). In this case the CPA system's compressor was adjusted to provide the highest energy output of the NIR OPA. Due to the lack of dispersion control of the OPA output, this fixed the pulse lengths of both the 1350 nm probe and 800 nm alignment pulse. The latter was measured to be approximately 75 fs at the interaction region via the standard SPIDER apparatus.

Overlapping alignment and probe beams

The procedure for spatially overlapping the alignment and probe beam was the same for 800 nm and 1350 nm probe pulses. It consisted of three steps. First, the alignment pulse was blocked and the probe pulse aligned through the interaction chamber. In the remaining two steps, the alignment beam was then aligned onto the probe beam, such that they co-propagated collinearly.

For the second step, a mirror was placed in front of the focussing lens L3 to pick up the alignment and probe beam. For convenience, the mirror was mounted on a magnetic base so it could easily be taken in and out of the beam line. Both beams then propagated along the imaging line displayed in the main setup photograph. In this way, the overlap of the beams could be monitored on a card before the mirror and at the end of the imaging line. Due to the large distance between those two points, a very accurate overlap of the beams could be achieved. The accuracy of the overlap was improved by using irises I2 and I3 for imprinting interference patterns on both beams. In particular, the probe beam was reduced to a single bright spot and the alignment beam to an interference pattern with a few rings. The BC and the preceding mirror in the alignment beam line were then used to center the alignment beam pattern onto the probe beam spot at the start and end of the imaging line.

Thirdly, the BC was fine adjusted to overlap the foci of the alignment and probe beams. The foci were imaged on a CCD camera (C) (WinCamD, Dataray) via the imaging line displayed in the inset of Fig. 4.9. For this a wedge in a flip mount (W1) was placed after the focussing lens. When flipped into the beam line, the focussing beam was picked up and directed to the camera via a second wedge (W2). Wedges were used to significantly reduce the intensity of the sampled beam, as only approximately 10 % of the pulse energy was reflected per wedge.

With the camera in the focal plane of the probe, the divergence of the alignment beam was adjusted by translating L2. In this way, the focal plane of the alignment pulse was placed onto the focal plane of the probe pulse. This was necessary for accounting for differences in size, divergence and wavelength (in the case of the 1350 nm probe) between the probe and alignment beams. Typical focal spots recorded via this procedure are displayed in Fig. 4.16 in Sect. 4.5.1.

The temporal overlap was facilitated by focussing the spatially overlapped beams in air. With both pulses overlapping in time, the resulting plasma breakdown was enhanced nonlinearly, which was easily visible even at moderate intensities. This procedure was suitable for determining the temporal overlap with an uncertainty of ± 0.08 ps, which was half of the temporal width of the visible plasma breakdown enhancement.

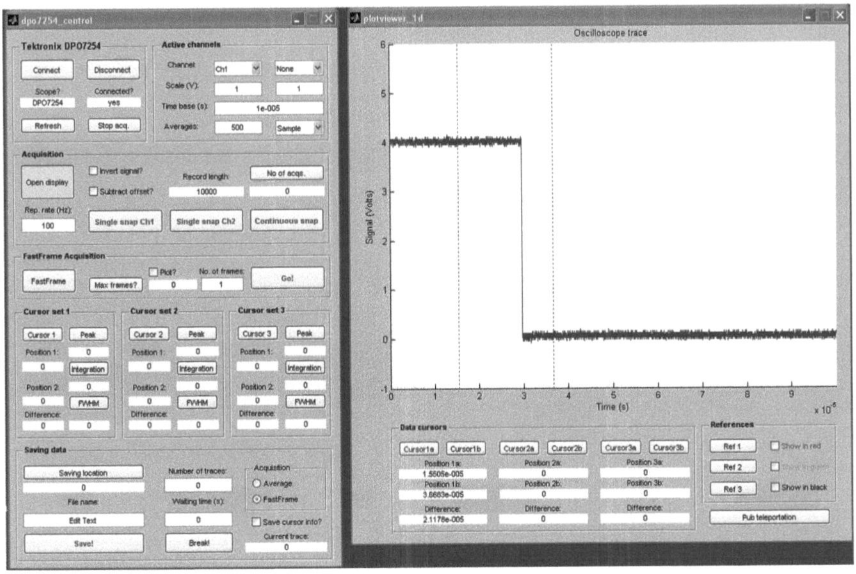

Fig. 4.10 Graphical user interfaces (GUIs) of the software components written for controlling the experimental setup. All programs were written in MATLAB. 'dpo7254_control' and 'plotviewer_1d' are the GUIs for the oscilloscope software. Note that cursors can be placed in the acquired traces for basic online analysis

4.3.3 Data Acquisition

For the experiments in this thesis, TOF mass spectra had to be recorded as a function of many different laser parameters, such as probe intensity and ellipticity, alignment pulse polarisation and temporal delay between alignment and probe pulse. In order to perform such scans efficiently and in a reproducible manner, the wave plates (QWP, HWP1, HWP2) and the translation stage (S) were motorised and software controlled. For synchronising the data acquisition with the scans, the oscilloscope that was used for recording the TOF spectra was software controlled as well. The software was written in MATLAB and the graphical user interfaces are displayed in Figs. 4.10 and 4.11.

The control softwares were linked via an additional control panel ('daq_tof2' in Fig. 4.11), where the scan parameters could be chosen and the settings sent to the control software of each individual component. In the following, the control panel will be explained briefly to present the possible scan parameter choices.

Scan type The choice of the scan type determines which laser parameter is scanned and which parameters are kept fixed. In the figure, 'Delay scan' is selected, meaning that the temporal delay between the pulses is scanned, while the wave plates are kept fixed. When changing the scan type, the assignment and units of the scan parameters

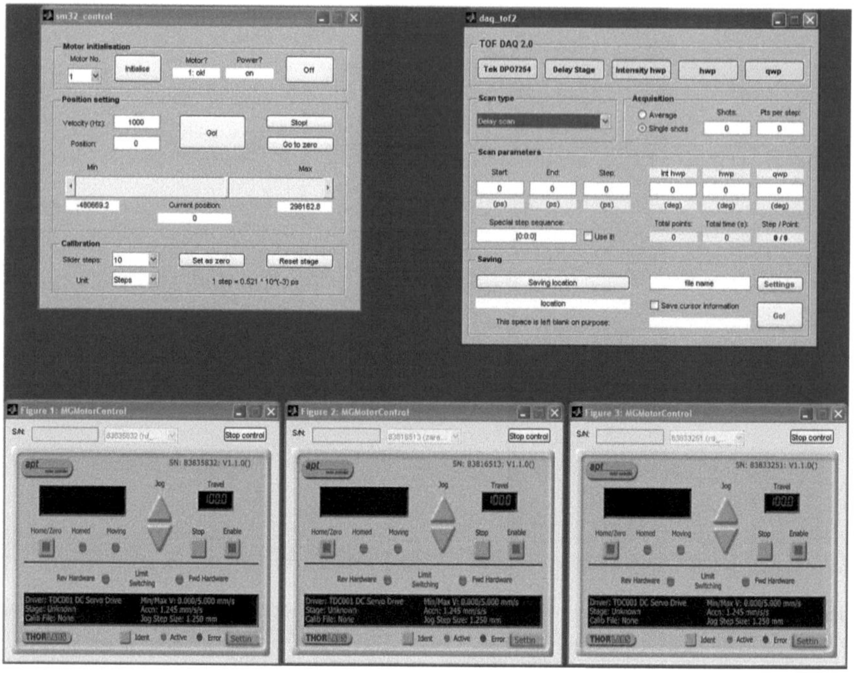

Fig. 4.11 Graphical user interfaces (GUIs) of the software components written for controlling the experimental setup. All programs were written in MATLAB. 'sm32_control' is the GUI for controlling the delay stage. 'MGMotorControl' is the software for driving the wave plate rotation stage and was provided by the manufacturer. It is loaded into MATLAB for each utilised roation stage (from *left* to *right*: HWP1, HWP2, QWP). On the *upper right* hand side, the GUI 'daq_tof2' is displayed that is used for linking the previous GUIs. It is described in more detail in the text

adjust accordingly in the control panel. Each laser parameter can be selected to be scanned.

Acquisition Two acquisition modes are possible: acquisition of a TOF spectrum that is averaged over several laser shots by the oscilloscope ('Average') or the acquisition of individual TOF spectra per laser shot ('Single shots'). In the first case the oscilloscope trace is transferred to the computer when sufficient individual traces were averaged. In the second case individual traces are saved on the oscilloscope at the trigger rate received by the oscilloscope. When the set number of traces is reached, all individual traces are transferred to the computer in form of a large data matrix. Furthermore the number of recorded laser shots ('Shots') and the number of acquisition iterations per scan step ('Pts. per step') can be chosen.

Scan parameters For equidistant scan steps, one may set the start and end value of the parameter scan and the step size on the left hand side of this panel. For scans with varying step size, a special step sequence may be entered as a vector containing the desired values of the scan. On the right hand side the values for the fixed scan parameters can be entered.

Saving File name and saving location can be assigned. 'Saving cursor information' refers to the possibility of placing cursors on the TOF spectra acquired via the oscilloscope control software. In this way information about the acquired curves in the cursor intervals is saved during the scan and can be accessed upon its completion. The 'Settings' button sends the scan parameters to the respective control softwares that position the motors accordingly. The 'Go' button then starts the scan.

4.3.4 Gas Mixing Setup

In order to deliver binary gas mixes to the interaction region, a setup for filling a sample cylinder (stainless steel, 0.5 l volume) with two different gases was built. A schematic drawing is displayed in part (a) of Fig. 4.12. The cylinder could be sealed via a needle valve and plugged into the gas mixing line via a quick release connection. The line had three inlets; one for each of the gases to be mixed and an evacuation line leading to a scroll pump (SH-110, Varian). The line further contained a pressure release valve at approximately 30 bar, a pressure sensor (PTX 600, Druck) and a check valve for preventing gas to flow back into the gas supplies and causing contamination.

For all presented experiments, gas 1 was a molecular sample; N_2, O_2 or CO_2. Gas 1 was the minor component in the mixture with gas 2, which was chosen to be Ar. The mixing procedure was as follows. With the cylinder open and connected to the mixing line, all lines were evacuated. The cylinder and gas line were then filled with gas 1 at a pressure p_1. The cylinder and the gas 1 inlet were closed and the mixing line evacuated. Next, the mixing line was filled with gas 2 at a pressure $p_2 > p_1$. Afterwards the cylinder was opened, such that the sample inside was topped up with gas 2, reaching the total pressure p_2. The cylinder was closed and removed from the mixing line. The mixing ratio of the gases was then assumed to be given by the ratio of their partial pressures $R = p_1/p_2$. The relative error of the pressure sensor measurement was 0.08 % according to the manufacturer. The utilised pressure reader, however, displayed the pressure in steps of 1 psi (1 bar $\approx$ 14.5 psi). The reading error was therefore estimated to be $\Delta p = 0.5$ psi. For pressure measurements below 625 psi, the resulting relative error fulfilled $\Delta p/p > 0.08\,\%$. As all measured pressures were below 300 psi, Δp was chosen as the total error of the pressure measurement. This resulted in $\Delta R = |R\Delta p|\sqrt{p_1^{-2} + p_2^{-2}}$ via standard error propagation.

The experiments presented in this thesis relied on the homogeneity of the gas mix. First of all, results needed to be reproducible, which means that the concentration of each component in the mix must remain constant over many laser shots. Secondly, the properties of the gas expansion depended on the partial pressures of the components in the mix. A variation of the mixing ratio between gas pulses would thus lead to significant changes in the supersonic expansion. It is not ensured, however, that inert

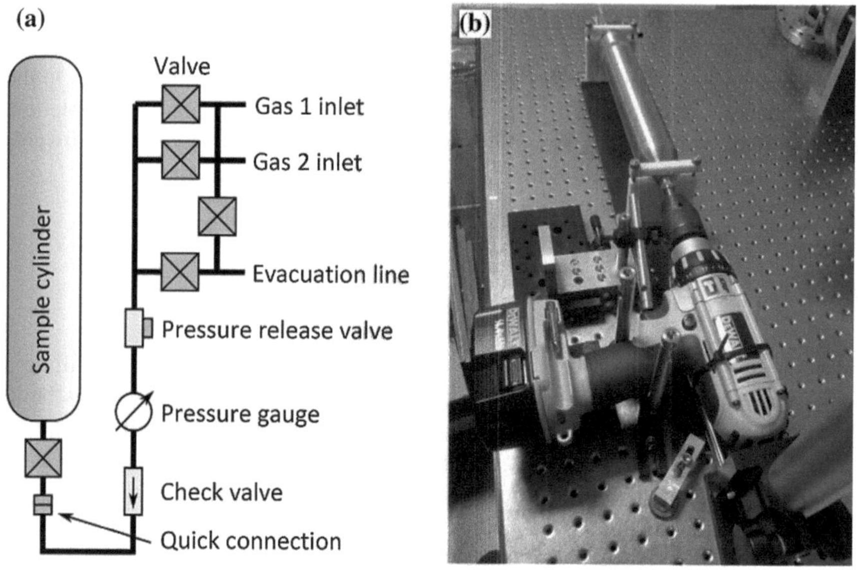

Fig. 4.12 Part **a** displays a schematic drawing of the utilised gas mixing line described in the text. Part **b** shows the setup for rotating the sample cylinder containing the gas mix with a drill. As the cylinder contained several steel balls, a homogeneous mixing process could be facilitated

gases form a homogeneous mixture when placed in the same container. Noble gases, for example, do not mix unless this is facilitated via an additional mixing setup.

For achieving a homogeneous binary gas mix for the experiments presented here, the cylinder was filled with about 20 steel balls of various diameters. The cylinder filled with the mixed sample was then mounted into the setup displayed in part (b) of Fig. 4.12, where it was connected to a battery powered drill (XRP, DeWalt). The cylinder was then rotated at a fixed speed for 30 min. Mixing of the gases was ensured due to the presence of the steel balls in the cylinder. After the mixing the sample cylinder was connected to the gas input line of the gas valve in the source chamber. A filter of 50 μm mesh size was placed in the gas line between the cylinder and the gas input line to avoid possible metal filings from the steel balls to clog the nozzle of the gas valve.

4.4 Molecular Beam

Molecular beams are an ideal tool for delivering gas samples for the study of laser-molecule interactions. This is because they are well localised, collimated, high density gas beams consisting of translationally and rotationally cool molecules. In this section the underlying gas dynamics are presented and the molecular beam used for the experiments in this thesis is characterised. The following discussion is based on the review by Scoles [37].

4.4.1 Supersonic Gas Jets

Molecular beams are extracted from supersonic gas expansions. Consider a gas placed in a reservoir at rest, backing pressure P_0 and temperature T_0. If the reservoir is connected to an evacuated chamber with background pressure $P_b \ll P_0$ via a short converging nozzle, the gas will rapidly expand when flowing from the reservoir into the vacuum chamber. This is because the gas must adapt to the boundary conditions in the vacuum chamber imposed by the much lower background pressure.

The resulting gas flow dynamics depend on the gas velocity at different stages of the expansion. It is conveniently expressed in form of the Mach number M that is the ratio between the velocity of the gas medium and the sonic velocity through it [38]. Starting at subsonic velocity ($M \ll 1$), the gas accelerates when exiting through the converging nozzle, due to the pressure gradient and the flow area reduction. The gas may reach sonic velocity ($M = 1$) at the nozzle exit if the pressure gradient is sufficiently large, such that:

$$P_0/P_b > 2.1 \tag{4.6}$$

This is an essential condition as the flow dynamics of the gas expansion outside the nozzle change with the Mach number. If $P_0 \approx P_b$, the gas exits with subsonic velocity and without substantial density and velocity changes. If Eq. 4.6 is fulfilled, however, the gas exits with supersonic velocity ($M > 1$) and due to the large pressure difference the gas density decreases rapidly during the expansion, which leads to further acceleration ($M \gg 1$). For this, the exiting gas flow is called a *supersonic gas jet*. Under these conditions, the flow field properties along the centreline of the expansion turn out to be almost completely independent of the thermodynamic gas properties:

$$M(x) = A\left(\frac{x}{d} - B\right)^{\gamma-1} - \frac{\frac{1}{2}\frac{\gamma+1}{\gamma-1}}{A\left(\frac{x}{d} - B\right)^{\gamma-1}} \tag{4.7}$$

where x is the distance from the nozzle exit, d the nozzle exit diameter and $\gamma = c_p/c_v$ the ratio between the specific heat and constant pressure and volume, respectively. A and B are empirical constants that depend on γ.

The thermodynamic properties of the expanding gas, however, depend on the flow field properties. For their analysis one assumes the expansion to be isentropic, which neglects viscous and heat conduction effects. With these energy loss mechanisms excluded, the thermodynamic properties during the expansion are directly related to the initial thermodynamic properties with the gas at rest in the reservoir, called the *stagnation state*. Here, the gas temperature T and its density n are the most relevant:

$$T(x) = T_0\left(1 + \frac{\gamma - 1}{2}M(x)^2\right)^{-1} \tag{4.8}$$

$$n(x) = n_0\left(1 + \frac{\gamma - 1}{2}M(x)^2\right)^{-1/(\gamma-1)} \tag{4.9}$$

where T_0 and n_0 are the corresponding stagnation state values.

4.4.2 Cooling Mechanism

During a supersonic expansion, the gas is accelerated at the expense of its internal energy. The resulting velocity distribution is thus shifted to a higher average value whilst its initial room temperature distribution is substantially narrowed to several Kelvin. This effect is referred to as *translational cooling*. Thermal equilibrium between rotational and vibrational degrees of freedom during the expansion is established via molecular binary collisions. In this way, the translational cooling induces rotational and vibrational cooling. This process is thus most efficient close to the nozzle exit where there is a high gas density and thus high collision frequency. With increasing distance from the nozzle, the jet density decreases. Eventually, the collision frequency will thus not be high enough to maintain thermal equilibrium between all degrees of freedom. This transition region is also referred to as the *freezing region*, because the temperatures and associated thermal properties of the degrees of freedom are frozen in. In the case of diatomic molecules, thermal equilibrium of vibrational degrees of freedom requires a total number of collisions on the order of 10^4. For rotational cooling on the other hand, 10^1 to 10^2 collisions are sufficient. For this, vibrational cooling freezes much earlier than rotational cooling.

The cooling mechanism is quantitatively described via the so-called *quitting surface model*. It divides the gas expansion into two zones via the freezing region for the translational temperature: (1) an isentropic continuum flow regime close to the nozzle where thermal equilibrium is maintained through binary collisions and (2) a collisionless, free molecular flow regime, where the terminal gas velocity is reached and the translational temperature thus remains constant.

Within this model, the translational temperature is separated into a parallel and perpendicular temperature, associated with the velocity distribution perpendicular and parallel to the gas propagation direction, respectively. In zone 1, thermal equilibrium is ensured, such that $T_{\parallel} = T_{\perp}$. In zone 2, $T_{\parallel} = const.$, yet $T_{\perp}$ may decrease further due to molecules with larger perpendicular momentum being driven away from the centreline of the gas expansion. This effect is called *geometrical cooling*. The quitting surface model is graphically illustrated in Fig. 4.13.

The location of the quitting surface x_q can be determined from the terminal Mach number M_T that is related to the terminal parallel velocity and its thermal distribution. For monoatomic gases (except He), it can be approximated as follows [39]:

$$M_T = 3.32(P_0[\text{bar}] \times d[\mu\text{m}])^{0.4} \tag{4.10}$$

The location of the quitting surface can then be obtained via Eq. 4.7:

$$M_T = M(x_q) \tag{4.11}$$

In practice, the binary collision rates $\Gamma^{(2)}$ during the jet expansion depend on the stagnation pressure P_0, the nozzle diameter d and the stagnation temperature T_0:

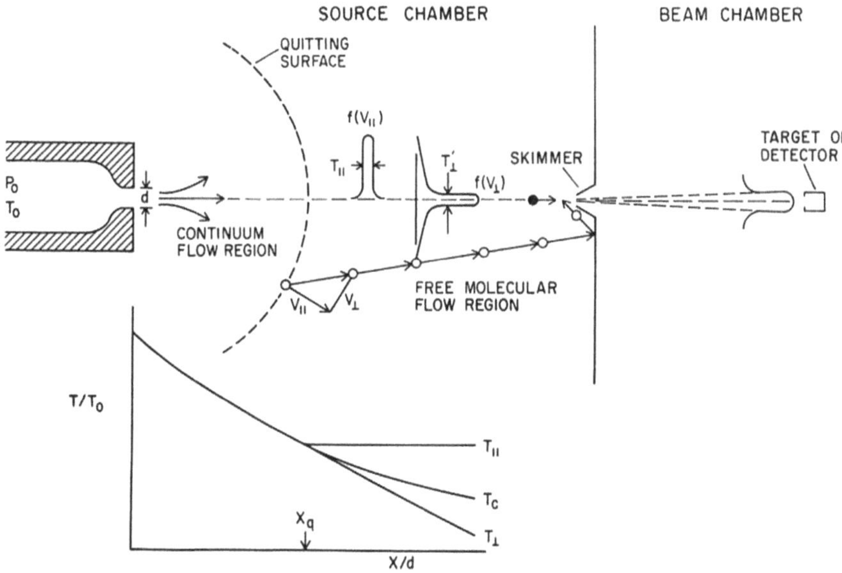

Fig. 4.13 Illustration of the quitting surface model. It displays the transition from continuum to free molecular flow. The associated zones are separated by the quitting surface located at x_q. From there onwards, thermal equilibrium between parallel and perpendicular temperatures cannot be maintained and $T_{\parallel}$ is frozen out, whilst $T_{\perp}$ decreases further due to geometrical cooling as shown in the temperature plot. T_c is the continuum temperature (Reprinted and adapted with permission from [37]. Copyright (1988) by Oxford University Press, Inc)

$$\Gamma^{(2)} \propto P_0 d/T_0 \tag{4.12}$$

For achieving a high degree of cooling, one thus needs to establish a high density gas flow through a large nozzle, which implies reaching a high terminal Mach number. This scaling is limited by several factors. Firstly, the gas flow rate scales as $P_0 d^2$ and requires sufficient vacuum pumping speed in the expansion chamber. If pumping is too low, the background pressure P_b becomes too high. This may lead to Eq. 4.6 not being fulfilled anymore or secondary collisions between gas jet molecules and background molecules or jet molecules scattered off the vacuum chamber walls. The latter would lead to re-heating of the jet.

Secondly, with increasing collision rates, three-body collisions become more likely as well. They lead to cluster formation, which is highly undesired if molecules in gas phase are studied. Three-body collision rates scale as:

$$\Gamma^{(3)} \propto P_0^2 d/T_0^2 \tag{4.13}$$

Usually, the Hagena scaling parameter Γ^* is used to estimate whether clustering in the jet is likely [40]:

$$\Gamma^* = kP_0[mbar](d[mm])^{0.85}(T_0[K])^{-2.29} \tag{4.14}$$

where k is an empirical condensation parameter depending on the molecular or atomic species. Clustering is expected to set in for $\Gamma^* > 300$.

Thirdly, the achievable cooling and onset of clustering depend on the molecular or atomic species undergoing the expansion. In the case of noble gases it is well known that He does not form clusters in typical supersonic jets [37, 41]. This is because He dimers are unstable and thus trimers are the smallest available cluster size, which only form at very low temperatures on the milliKelvin level. Going from high to low temperature, Xe clusters first, followed by the other noble gases in order of decreasing atomic weight. This is reflected in their respective condensation parameter which is $k_{Ar} = 1650$ for Ar and $k_{He} = 3.85$ for He, for example [42].

4.4.3 Skimmed Molecular Beams

Skimming the supersonic gas jet has several advantages and is illustrated in Fig. 4.13 Firstly, one may select the central and coolest part of the jet and thus extract a narrow, collimated molecular beam. Secondly, the vacuum chamber is divided into a source and a beam chamber, which significantly reduces the gas load in the beam chamber. This effect is usually referred to as differential pumping and has been discussed in Sect. 4.3.1.

A skimmer attenuates the gas density $n(x_f)$ at a final distance x_f from the nozzle source. The transmission efficiency η of a skimmer with an aperture radius r, located at a distance x_s from the nozzle is given as:

$$\eta = 1 - \exp\left[-\frac{2M_T^2}{\gamma}\left(\frac{r}{x_q}\right)^2\left(\frac{x_f}{x_f - x_s}\right)^2\right] \tag{4.15}$$

The molecular beam density m at x_f is then obtained via Eq. 4.9, such that:

$$m(x_f) = \eta n(x_f) \tag{4.16}$$

Placing a skimmer in the gas expansion chamber may also have detrimental effects on the molecular beam. The most important one is the possibility of gas molecules being scattered of the skimmer wall back into the gas jet, as this may lead to a re-heating of the transmitted molecules or pushing jet molecules away from the centre line and thus not being transmitted. This is illustrated in Fig. 4.13. In addition to such skimmer interference effects, shock waves within the gas expansion may be formed in front of the skimmer if the background pressure is too high ($P_b > 10^{-4}$ mbar). For avoiding these effects, different skimmer geometries have been investigated [43]. Generally, however, maintaining a low background pressure during gas jet operation reduces the above mentioned effects.

Table 4.1 Parameters used to calculate the properties of the molecular beam used in the experiments presented in this thesis

γ	d (μm)	r (mm)	x_s (mm)	x_f (mm)	P_0 (bar)	P_b (mbar)	P_i (mbar)
5/3	300	0.5	135	369	2.2	2×10^{-5}	6×10^{-8}

Note that P_i represents the background pressure in the interaction chamber

4.4.4 Seeded Gas Jets

Seeding a molecular species in a light carrier gas may lead to enhanced cooling of the seed [37, 41]. The reason for this is that lighter species are more easily accelerated and hence reach higher terminal Mach numbers. As a result, the seeded species may reach higher terminal Mach numbers than in a pure expansion. In the continuum flow regime, collisions between seed and carrier establish thermal equilibrium, which leads to the enhanced cooling of the seed, given that the carrier may achieve lower translational temperatures. This effect has been demonstrated for various seed species, including large organic molecules [44]. For light species seeded in a heavier carrier gas, it has been found that the translational temperature of the seed can even be lower than that of the carrier [45].

As discussed before, the cooling of the seed is limited by condensation in the gas jet. For this, He is typically used as a carrier gas. For the experiments presented in this thesis Ar was used as a carrier gas, due to having comparable ionisation potentials to the utilised seed molecules. This was used for a more accurate interpretation of the acquired data.

4.4.5 Molecular Beam Characterisation

In this section the molecular beam used for the experiments is characterised. The experimental parameters used in the calculations are displayed in Table 4.1 and represent typical conditions from the conducted experiments. Ar was used as a gas species, which required the use of the parameter values $A = 3.26$ and $B = 0.075$ in Eq. 4.7 (see [37]). The gas density in the stagnation state was calculated via the ideal gas law, resulting in:

$$n = \frac{P[bar] \times 10^5}{kT} \tag{4.17}$$

where P is the pressure in bar, k the Boltzmann constant and T the temperature. n_0 was obtained by plugging in the value for P_0 and assuming room temperature, such that $T_0 = 293\,\text{K}$.

The previously discussed equations were then used to calculate the quantities displayed in Table 4.2. First of all, the low background pressure in the source chamber ensures a high ratio of $P_0/P_b \gg 2.1$. Condensation in the resulting supersonic

Table 4.2 Obtained results for the characterisation of the molecular beam

P_0/P_b	M_T	x_q (mm)	n_0 (cm^{-3})	$n(x_f)$ (cm^{-3})	η	$m(x_f)$ (cm^{-3})	Γ^*
1×10^8	45	15	5×10^{19}	5×10^{12}	1	5×10^{12}	3

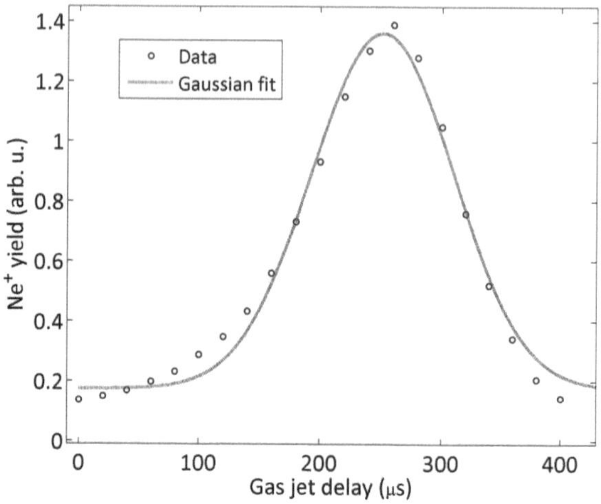

Fig. 4.14 Ne^+ signal as a function of delay between the gas jet trigger and the laser pulse, mapping out the temporal molecular beam profile. For the measurement, 800 nm, 30 fs at $4 \pm 1 \times 10^{14}$ Wcm^{-2} pulses were used. The gas jet was operated at $P_0 = 1$ bar with an opening time of 150 μs set on the delay generator

expansion, however, is unlikely due to $\Gamma^* \ll 300$. The most important result is the estimate of the gas density in the interaction region given by $m(x_f) = 5 \times 10^{12}$ cm^{-3}. Compared to the background gas density of $n_i = 1 \times 10^9$ cm^{-3}, this ensures a high contrast of target density in the interaction region.

The gas jet was operated in pulsed mode at a repetition rate of 47 Hz. This ensured low background pressures in the interaction chamber. The temporal profile of the gas pulse was measured by recording the ionisation signal as a function of time delay between the gas valve trigger and the laser pulse. The resulting curve is displayed in Fig. 4.14. The extracted FWHM is 138 μs.

The spatial profile of the molecular beam was measured by translating the laser focus relative to the gas jet and recording the resulting Ar^+ ionisation rate. The focus was translated by moving the focussing lens. Figure 4.15 displays the recorded spatial profiles along the x- and z-axis. The x-direction corresponds to translating the focus perpendicular to the molecular beam and parallel to the TOF axis, wheras the z-direction corresponds to the laser propagation direction. From the Gaussian fits displayed in the figure, the FWHM was extracted for both profiles, resulting in $\Delta_x = 2.7$ mm and $\Delta_z = 5.4$ mm. Note that $\Delta_z > \Delta_x$ because the ionisation signal obtained from translating the focus in z-direction is a convolution between the changing effective intensity at the interaction region and the molecular beam intensity. This effect can be neglected when moving the focus in x-direction as the focal spot diameter is much smaller than the molecular beam dimensions.

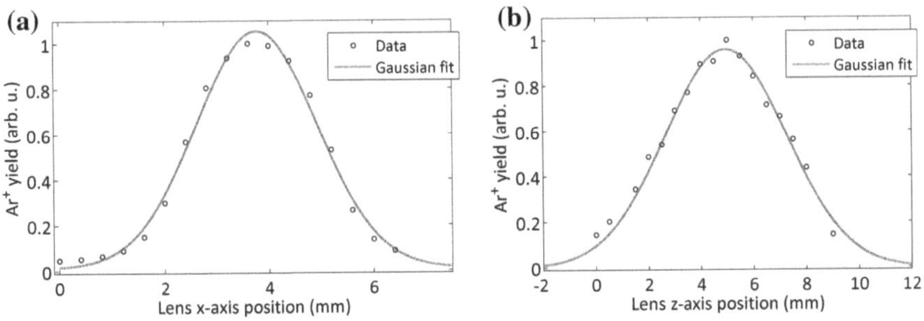

Fig. 4.15 Ar^+ signal as a function of lens position in x- (**a**) and z-direction (**b**), mapping out the spatial molecular beam profile. For the measurement, 800 nm, 30 fs at $4 \pm 1 \times 10^{14}$ Wcm^{-2} pulses were used. The gas jet was operated at $P_0 = 1$ bar with an opening time of 150 μs set on the delay generator

As the molecular beam is cylindrically symmetric around its axis of propagation, Δ_x represents the diameter of the jet at the interaction region. With focal spot diameters at FWHM of $\Delta w^{(800)} = 41\,\mu m$ and $\Delta w^{(1350)} = 61\,\mu m$ (see Table 4.3) one can thus estimate the number of target atoms or molecules n_t in the associated interaction volumes. This results in $n_t^{(800)} = 2 \times 10^7$ and $n_t^{(1350)} = 4 \times 10^7$.

4.5 Manipulation of Laser Pulse Parameters

In this section the manipulation of the laser parameters within the experimental setup is discussed in more detail.

4.5.1 Focussing

The laser beam focus produced by a lens is characterised by the unfocussed beam's waist radius W, wavelength λ and the lens' focal length f. Assuming a Gaussian intensity profile and zero divergence of the laser beam, the minimum radius w of the produced focus is given by [46]:

$$w = \frac{f\lambda}{\pi W} \tag{4.18}$$

The divergence of the focus is given by the Rayleigh range:

$$z_R = \frac{\pi w^2}{\lambda} \tag{4.19}$$

Table 4.3 Measured focal spot diameters for the different beam line arrangements

	800/800: probe	800/800: alignment	1350/800: probe	1350/800: alignment
$2w$ (μm)	70 ± 7	97 ± 8	103 ± 7	90 ± 7
Δw (μm)	41 ± 4	57 ± 5	61 ± 4	53 ± 4
z_c (mm)	9.6 ± 3.8	18.5 ± 5.3	12.4 ± 4.8	15.9 ± 4.9

Here, 800/800 refers to 800 nm alignment and probe pulses, whilst 1350/800 refers to 1350 nm probe and 800 nm alignment pulse. The confocal parameter is calculated from the value for w and the associated error is obtained via standard error propagation from the error estimated for w

It is associated with the distance over which the cross-sectional area of the focussed beam doubles. The confocal parameter $z_c = 2z_R$ is often used to estimate the length of the focus in the direction of propagation. Table 4.3 displays typical values for experimentally determined focal spot diameters at $1/e^2$ ($2w$), FWHM (Δw) and the resulting confocal parameter for the beamline arrangements used in the experiments. Here, the diameters at different intensity thresholds are related via $2w = (\sqrt{2}/\sqrt{\ln 2})\Delta w$.

Note that measuring the focal spot diameter in the case of the 1350 nm probe pulse requires two-photon absorption on the camera's CCD chip. For this, the measured focal spot corresponds to the squared intensity profile. In the values presented in Table 4.3, this is taken into account by multiplying the measured focal spot diameter by $\sqrt{2}$. Furthermore, one may expect that this effect removes low intensity features from the recorded beam profile. This suggests that the measured spot size underestimates the real spot size. This was taken into account by increasing the measured focal spot diameter $\Delta m'$ by 10 %, such that the real spot size is estimated as $\Delta w = \sqrt{2}(1.1\Delta w')$

In order to assess the quality of the probe focus, its unfocussed beam radius was determined via a knife-edge measurement before the focussing lens. For this a razorblade was inserted into the beam and the power of the clipped beam measured as a function of blade insertion. Mathematically, the measured power is then given as (see for example [47]):

$$P(x) = \frac{P_{tot}}{2}\left[1 - \text{erf}\left(\frac{x - x_0}{W_x}\right)\right] \tag{4.20}$$

where x denotes the direction of the blade insertion, x_0 its arbitrary initial insertion, W_x the beam radius at $1/e^2$ in x direction and P_{tot} the total power in the beam. Fitting Eq. 4.20 to the recorded data, $W_x = 3.4$ mm at $1/e^2$ was obtained. This beam was then focussed via a 30 cm focal length lens and the expected focal spot radius calculated via Eq. 4.16 using the extracted value of W_x and $\lambda = 800$ nm. This resulted in $w_x^{th} = 22\,\mu$m at $1/e^2$. From these values the M^2 value of the overall setup can be calculated:

$$M^2 = \left(\frac{w_x^{(exp)}}{w_x^{(th)}}\right) \tag{4.21}$$

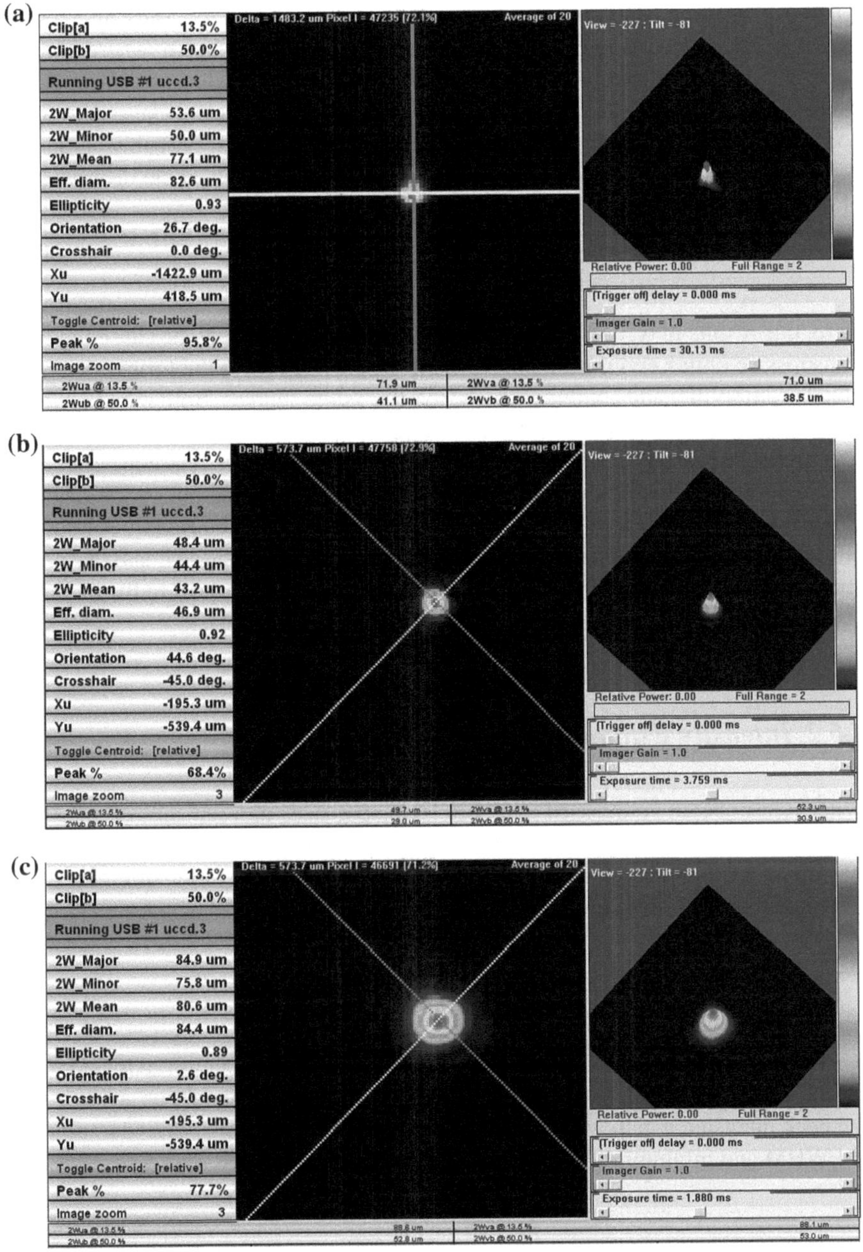

Fig. 4.16 Focal spot measurements conducted via a CCD camera and the commercially available software Dataray. The figure displays a typical focal spot of the 800 nm probe pulse (**a**), the 1350 nm probe pulse (**b**) and the 800 nm alignment pulse **c** obtained from the beamline arrangement employing 1350 nm probe pulses

where $w_x^{(exp)}$ is the experimentally measured beam radius at $1/e^2$. Here, this results in $M^2 = 2.5$. The parameter quantifies the quality of the laser beam, where $M^2 = 1$ corresponds to a diffraction limited Gaussian beam. Higher values are related to an overall divergence of the propagating beam and possible higher Gaussian modes contained in the beam profile. In the experiment this was most probably caused by the clipping of the beam at optics of insufficient size. The half-angle divergence α of the beam is then given as:

$$\alpha = M^2 \frac{\lambda}{\pi w_x^{(exp)}} \tag{4.22}$$

Here, this results in $\alpha = 1.1°$. Such a significant divergence may be caused by bent transmissive optics or a misalignment of telescope setups. For the former point, the glued beam splitter and combiner are the main possible sources. Both were, however, tested prior to use in the setup and did not induce divergence. For this, the divergence was probably caused by a slight misalignment of the reflective telescope employed for enlarging the beam radius of the CPA system close to its output. Additionally, a larger focal spot compared to the diffraction limit can also be caused by aberrations introduced by the lens. Here, the most important ones are spherical and chromatic aberrations as well as slight misalignments of the lens with respect to the incoming laser beam.

4.5.2 *Intensity Estimation*

When a laser beam is focussed tightly to focal spot sizes of a few tens of micrometers, the resulting intensity distribution at the focus has a strong spatial dependence. Assuming a Gaussian beam profile perpendicular to the propagation direction and a Lorentzian profile along this direction, the intensity distribution is given as [13]:

$$I(r, z) = \left(\frac{I_0}{1 + (z/z_R)^2}\right) \exp\left\{\frac{-2r^2}{w^2[1 + (z/z_R)^2]}\right\} \tag{4.23}$$

where r denotes the perpendicular distance from the propagation axis and z the position on this axis relative to the position of the waist w of the focus at $1/e^2$. I_0 is the peak intensity, which was estimated via the laser pulse energy ΔE, duration $\Delta\tau$ at FWHM and minimum focal spot diameter at FWHM Δw as:

$$I_0 = \frac{\Delta E}{\Delta\tau(\Delta w/2)^2\pi} \tag{4.24}$$

This estimate of the peak intensity was tested by conducting ionisation saturation measurements in Ar and N_2. Following the analysis developed in [48] the peak intensity at which ionisation saturation sets in is obtained from the x-axis intercept of a linear fit to the high intensity yields when plotted on a semi-log scale. This is

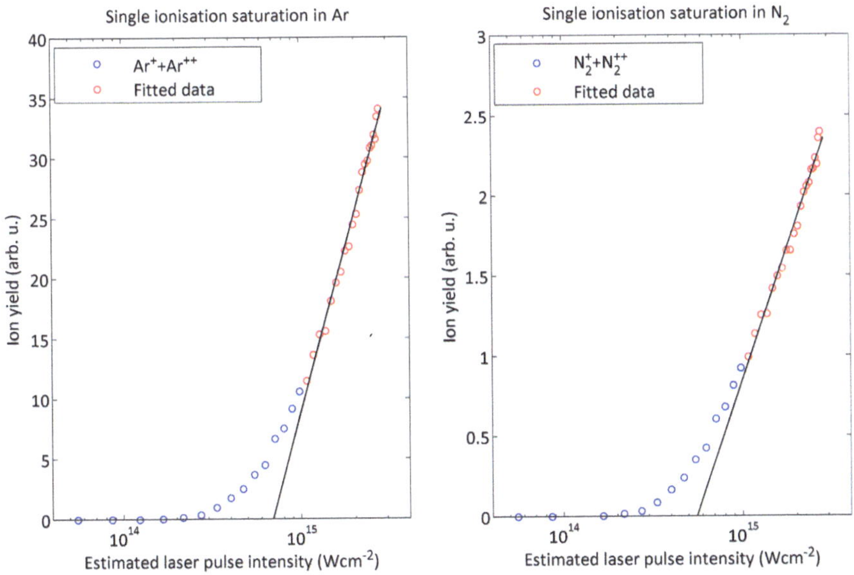

Fig. 4.17 Ionisation saturation measurements in Ar and N_2 via 800 nm, 30 fs laser pulses. Following the procedure explained in the text, the intensity axis intercept of a *linear* fit to the high intensity data points is associated with the saturation intensity I_{sat}

shown in Fig. 4.17 where the total single ionisation yield was calculated as the sum of singly and doubly charged ions. Higher charge states were neglected due to their marginal yields in the recorded intensity region. The saturation intensities in Ar and N_2 were found then to be $I_{sat}^{(Ar)} = 7 \times 10^{14}\,\mathrm{Wcm^{-2}}$ and $I_{sat}^{(N_2)} = 5 \times 10^{14}\,\mathrm{Wcm^{-2}}$. The measurement for Ar agrees very well with a previous measurement [49], where a saturation intensity of $7.5 \times 10^{14}\,\mathrm{Wcm^{-2}}$ was found. This verifies the peak intensity estimates for high intensities in the saturation regime, where low intensity processes will be masked. Unfortunately, reference measurements for N_2 were not available.

However, it has to be assumed that the peak intensity is not an adequate estimate for the effective intensity in the regime where the strong field experiments were conducted. This is because focal volume averageing in the interaction region will lead to a range of intensities contributing to the overall ion yields. For this, the average intensity I_{av} corresponding to a given peak intensity I_0 was estimated as follows. First, the spatial intensity distribution was calculated via Eq. 4.23. I_{av} was then determined as the average over all grid points for which $I(r, z) > 8 \times 10^{13}\,\mathrm{Wcm^{-2}}$ and $|z| < \Delta_x/2$. Here, $\Delta_x/2 = 1.4$ mm was the width of the molecular beam at FWHM.

The intensity threshold was chosen to account for the fact that at lower intensities the probability of the relevant strong field phenomena such as double ionisation and dissociation are signficantly reduced. When calculating the average intensity of a typical alignment pulse, the intensity threshold was set at $I(r, z) > 6 \times 10^{12}$ $\mathrm{Wcm^{-2}}$ assuming that rotational coupling is significantly reduced below this intensity

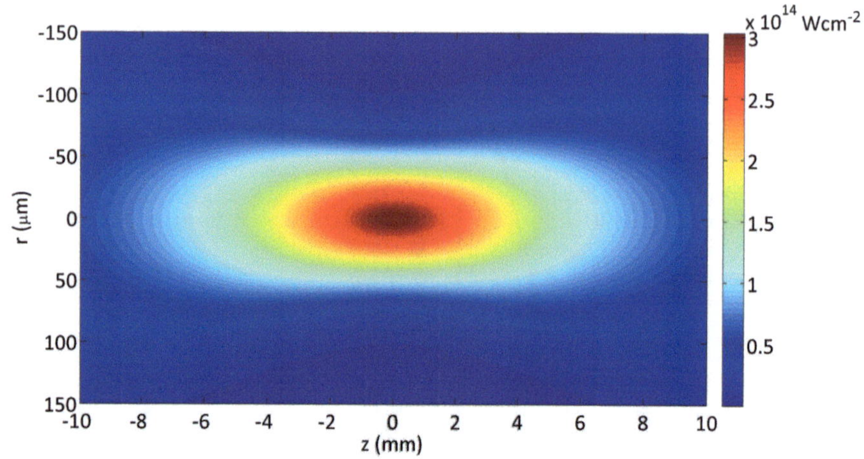

Fig. 4.18 Calculated spatial intensity distribution of a focussed laser pulse with the pulse and focussing parameters displayed in the first row of Table 4.4. The z-axis is the pulse propagation axis and r denotes the *perpendicular* distance to this axis

Table 4.4 Laser pulse and focal spot properties used to estimate the average intensity present in the interaction region in the experiments conducted for this thesis

λ (nm)	ΔE (μJ)	$\Delta\tau$ (fs)	Δw (μm)	w (μm)	z_R (μm)	I_0 (Wcm^{-2})	I_{av} (Wcm^{-2})
800	120	30	41	35	4.8	3.0×10^{14}	$(2.0 \pm 1.0) \times 10^{14}$
	190	30	41	35	4.8	4.8×10^{14}	$(2.9 \pm 1.9) \times 10^{14}$
	100	70	57	49	9.2	5.6×10^{13}	$(3.2 \pm 2.4) \times 10^{13}$
1350	330	30	61	51	6.2	3.8×10^{14}	$(2.4 \pm 1.4) \times 10^{14}$
	540	30	61	51	7.6	6.2×10^{14}	$(3.6 \pm 2.6) \times 10^{14}$

Note that the 800 nm, 70 fs case is associated with a typical alignment pulse. The remaining calculations correspond to probe pulses

compared to the higher intensities present in the pulse. The constraint on the distance was due to the interaction volume dimension along the laser propagation direction imposed by the gas jet width. All calculations employed a grid spacing of 0.05 μm in r- and 0.005 mm in z-direction. Figure 4.18 shows a calculation for the intensity of a focus obtained from an 800 nm probe pulse with $\Delta E = 120\,\mu$J, $\Delta\tau = 30$ fs and the focal spot dimensions displayed in Table 4.4. Table 4.4 then displays the average intensities calculated for typical experimental parameters that were employed in the experiments presented in this thesis.

One should point out, however, that the obtained average intensities are relatively crude estimates due to large number of input parameters for the calculation. For this, the error ΔI for I_{av} was estimated to be relatively large and determined as $\Delta I = I_0 - I_{av}$. Furthermore, when quoted in the following experimental chapters, the values are rounded to one significant digit only. This is done for suggesting a realistic precision of the determined value for the average intensity.

4.5.3 Polarisation State

Wave plates are made of birefringent materials, which have different indices of refraction along perpendicular axes in the crystal the plate is made of. For a wave plate, these axes are parallel to the face of the plate and thus normal to the incident beam. The principal axis of the plate is defined as the angular position (when rotated around its normal) that leaves a linearly polarised beam unaltered. When rotated away from the principal axis by an angle θ, the beam's electric field is split into perpendicular components oscillating along the crystal's axes. Due to the difference in refractive index, the components travel different optical path lengths, which leads to a phase shift ϕ between them. The result is a change of the polarisation state of the transmitted electric field.

Half-wave plate

Half-wave plates introduce a phase shift of $\phi = \pi$. This leaves the incident linear polarisation intact and rotates the polarisation plane by an angle of 2θ in the direction of of the wave plate rotation. This can be used for attenuating the intensity of a linearly polarised laser beam, by positioning a polariser after a half-wave plate. With the principal axis of the polariser aligned along the initial polarisation plane of the beam, the intensity transmitted through the attenuator follows Malus' law [50]:

$$I(\theta) = I_0 \cos^2(2\theta) \tag{4.25}$$

where I_0 is the intensity of the incident beam and θ the angle between the principal axes of the half-wave plate and the polariser.

Quarter-wave plate

Quarter-wave plates introduce a phase shift of $\phi = \pi/2$. This creates elliptical polarisation, with the resulting polarisation ellipse characterised by the plate rotation angle θ. The eccentricity ϵ of the ellipse, which is a measure of the ratio the electric field amplitudes along its major (E_m) and semi-major axis (E_{sm}), is given by:

$$\epsilon(\theta) = \tan(\theta) = \tan\left(\frac{E_{sm}}{E_m}\right) \tag{4.26}$$

ϵ is thus often used to denote the ellipticity of the laser beam. The orientation of the polarisation ellipse, however, is rotated such that the major axis of the ellipse is at an angle θ with respect the the beam's initial linear polarisation direction. In the special case of $\theta = \pi/4$, $E_{sm} = E_m$ and the transmitted beam is circularly polarised. This means that $\epsilon = 1$.

4.5.4 Calibration of Polarisation Optics

Several polarisation optics in the beamline (see Fig. 4.9) had to be calibrated. This included the halfwave plate (HWP1) and polariser (P) forming a variable attenuator, the quarter-wave plate (QWP) and halfwave plate (HWP2) when placed in the probe arm and the halfwaveplate (HWP2) when placed in the alignment arm. In this section the employed calibration procedures are presented.

Variable attenuator

Without any polarisation optics in the beam, a power meter was placed after the beam combiner (BC). The incoming beam was P-polarised, so the polariser (P) was placed in the beam such that the transmitted power recorded by the power meter was maximised. In particular, only the left-right tilt of the polariser was adjusted, where the maximum transmission corresponds to the Brewster angle. The up-down tilt was fixed by the assembly for mounting it on the optical table.

With the polariser in the beamline, HWP1 was placed before the polariser. It was calibrated by recording the transmitted power at the power meter as a function of the waveplate angle. Due to Malus' law in Eq. 4.25, a function of the form $P(x) = a\cos^2(2(x-b)) + c$ was fitted to the recorded power $P(x)$ as a function of waveplate angle x. Here, a gives the maximum transmitted pulse energy, b the waveplate angle for which its principal axis is parallel to the principal axis of the polariser and c the minimum transmitted pulse energy when the incoming laser polarisation is rotated to be perpendicular to the polariser's principal axis.

Figure 4.19 shows a typical calibration curve obtained via the above procedure. The transmitted pulse energy is plotted as a function of laser polarisation angle with respect to the principal axis of the polariser. The conversion to peak intensity was made via the measured FWHM of the focal spot diameter and pulse duration. The error bars are estimated from the fluctuations of the pulse energy measurement at the power meter.

Quarter-wave plate in probe arm

The calibration of the quarter-wave plate was done by recording the double ionisation yield of Ar as a function of quarter-wave plate angle. For this, pure Ar or a gas mix including Ar was used as a gas sample for the molecular beam. It is well known, that the double ionisation of Ar is predominantly recollision induced at a sufficiently low intensity of about 2×10^{14} Wcm^{-2}. This implies that the resulting Ar^{++} yield has a strong ellipticity dependence, which takes the form of a Gaussian. The principal axis of the quarter-wave plate was thus obtained by fitting a function of the form $A(x) = a\exp\left(-\frac{(x-b)^2}{c^2}\right) + d$ to the recorded Ar^{++} yield $A(x)$, where x is the quarter-wave plate angle.

A typical calibration curve is displayed in Fig. 4.20. For the corresponding measurement, TOF spectra were acquired in averageing mode with 500 shots per spectrum. 800 nm, 30 fs laser pulses were used at an average intensity of 2×10^{14} Wcm^{-2}. From the TOF spectra, the Ar^{+} and Ar^{++} yields were obtained by numerically integrating the corresponding peaks. The ion yield ratio Ar^{++}/Ar^{+} was then plotted as a

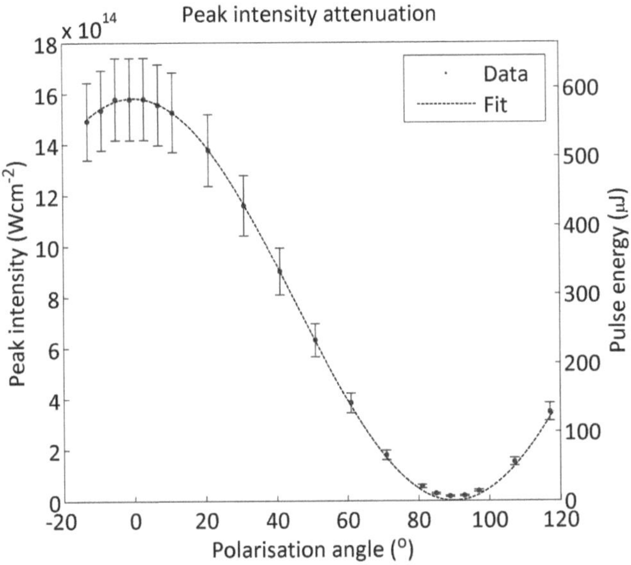

Fig. 4.19 Peak intensity resulting from the pulse energy transmitted through the variable attenuator in the beamline as a function of laser polarisation angle with respect to the principal axis of the polariser. Fitting Malus' law to the recorded data then provides a calibration of the attenuator relating the waveplate angle to the transmitted pulse energy

function of the quarter-wave plate angle and used for the fitting procedure described above. Using the ion yield ratio had the advantage of compensating for intensity fluctuations between data points, as the Ar^{+} variation as a function of ellipticity is much lower than the Ar^{++} variation. In the example displayed in Fig. 4.20 the Gaussian fit resulted in a position of the principal axis at a quarter-wave plate angle of 20.4°. This was the position of the plate that maintained the linear laser polarisation. The advantage of this calibration method is that it is done in situ under the employed experimental conditions.

Half-wave plate in alignment arm

When placed in the alignment arm, HWP2 had to be calibrated with respect to the probe polarisation. This was because the linear alignment pulse polarisation defined the orientation of the molecular sample in the laboratory frame and thus with respect to the probe. For the calibration, the angular tunneling ionisation rate of CO_2 was recorded as a function of halfwave plate angle and thus alignment pulse polarisation with respect to the linear probe pulse polarisation. Due to the broad nodal plane perpendicular to the molecular axis in the HOMO of CO_2, the tunneling ionisation rate is strongly supressed when the alignment pulse polarisation is perpendicular to the probe polarisation.

For the measurement, impulsive molecular alignment was induced by the alignment pulse and the probe delayed to interact with the molecular sample at the half

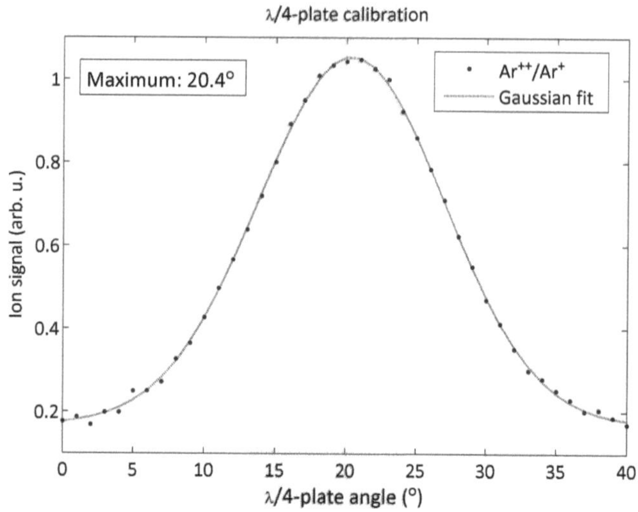

Fig. 4.20 Ion yield ratio Ar^{++}/Ar^{+} as a function of quarter-wave plate angle using 800 nm, 30 fs laser pulses at 2×10^{14} Wcm^{-2}. Due to the strong ellipticity dependence of Ar^{++} formation in this intensity regime, the principal axis of the quarter-wave plate can be extracted from a Gaussian fit to the data

revival. The CO_2^+ yield was then recorded as a function of halfwave plate angle. A typical calibration curve is displayed in Fig. 4.21. In this measurement a quarter-wave plate angle of 21.3° corresponded to linear probe polarisation. 800 nm, 30 fs pulses at 2×10^{14} Wcm^{-2} were used. TOF spectra were acquired in averageing mode with 500 shots per spectrum. The recorded data is displayed as blue data points. The halfwave plate angle corresponding to the minimum of the ion yield is obtained by fitting a parabola to the data. In this calibration, the alignment pulse polarisation was perpendicular with respect to the probe polarisation at a halfwave plate angle of 167°. This implies that the principal axis of the quarter-wave plate is parallel to the probe polarisation at 122°.

For measuring the ellipticity dependence of laser-molecule interactions in aligned molecules, the alignment pulse polarisation has to remain parallel to the major axis of the probe polarisation ellipse. However, when the quarter-wave plate is rotated by an angle ϕ, the major axis of the polarisation ellipse remains aligned along the principal axis of the quarter-wave plate. This implies that the major axis is rotated by the angle ϕ in the laboratory frame in the direction of the plate rotation. Therefore, the alignment pulse polarisation has to be rotated accordingly by the same angle. This requires a rotation of the halfwave plate by $\phi/2$. This was tested by a second calibration scan, where the quarter-wave plate was rotated by $\phi = 10°$ to a quarter-wave plate angle of 31.3° (red data points in Fig. 4.21). This corresponded to a probe pulse ellipticity of $\epsilon = 0.18$. Again, a parabola was fitted to the data to obtain the halfwave plate angle that rotates the alignment pulse polarisation to be perpendicular to the major axis of the probe polarisation ellipse. Here, this angle was found to be 162°, as expected.

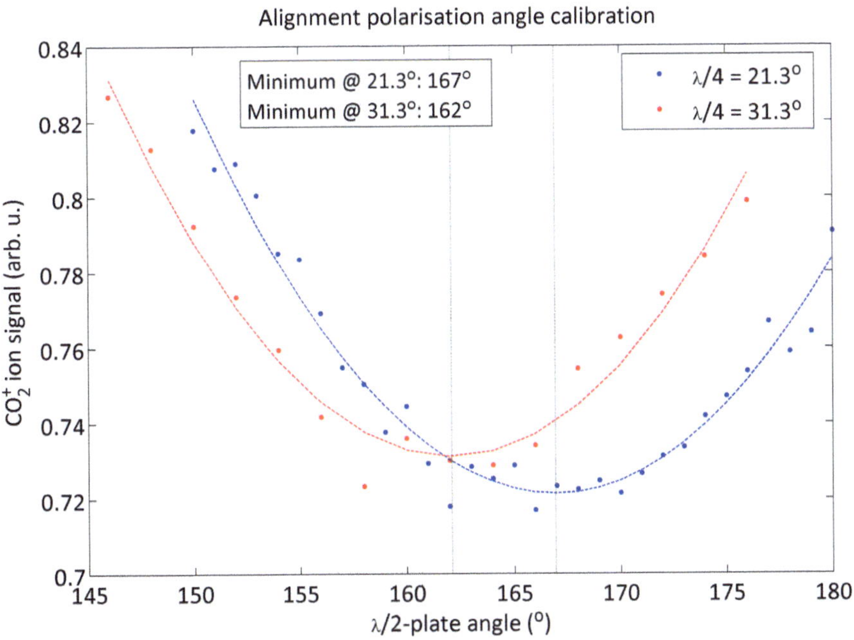

Fig. 4.21 CO_2^+ yield as a function of halfwave plate angle rotating the alignment pulse polarisation with respect to the probe pulse. The time delay between the pulses was set, such that the half revival is probed. 800 nm, 30 fs probe pulses were employed at 2×10^{14} Wcm^{-2} and two different ellipticities: $\epsilon = 0$ for a quarter-wave plate angle of $\phi = 21.3°$ (*blue data*) and $\epsilon = 0.18$ for $\phi = 31.3°$ (*red data*). For details on the calibration procedure see text

In addition, such a test measurement establishes the rotation direction of the halfwave plate for compensating a given rotation of the quarter-wave plate. In this example, increasing the quarter-wave plate angle ϕ has to be compensated by decreasing the halfwave plate angle θ, such that $\theta = -\phi/2$. This was due to the arrangement of the wave plate rotation stages in the beamline.

Half-wave plate in probe arm

In case the halfwave plate HWP2 is placed in the probe arm in place of the quarter-wave plate, the same calibration procedure was followed as with HWP2 placed in the alignment arm. The only difference was that in this case the probe polarisation was rotated with respect to the fixed alignment pulse polarisation.

References

1. M.N.R. Ashfold, N.H. Nahler, A.J. Orr-Ewing, O.P.J. Vieuxmaire, R.L. Toomes, T.N. Kitsopoulos, I.A. Garcia, D.A. Chestakov, S. Wu, D.H. Parker, Imaging the dynamics of gas phase reactions. Phys. Chem. Chem. Phys. **8**(1), 26 (2006)

2. M.N.R. Ashfold, J.D. Howe, Multiphoton spectroscopy of molecular species. Annu. Rev. Phys. Chem. **45**(1), 57–82 (1994)
3. A.I. Chichinin, K.H. Gericke, S. Kauczok, C. Maul, Imaging chemical reactions—3D velocity mapping. Int. Rev. Phys. Chem. **28**(4), 607 (2009)
4. D.W. Chandler, P.L. Houston, Two-dimensional imaging of state-selected photodissociation products detected by multiphoton ionization. J. Chem. Phys. **87**, 1445–1447 (1987)
5. M. Lampton, O. Siegmund, R. Raffanti, Delay line anodes for microchannel-plate spectrometers. Rev. Sci. Instrum. **58**, 2298–2305 (1987)
6. O. Jagutzki, V. Mergel, K. Ullmann-Pfleger, L. Spielberger, U. Spillmann, R. D/"orner, H. Schmidt-B/"ocking, A broad-application microchannel-plate detector system for advanced particle or photon detection tasks: large area imaging, precise multi-hit timing information and high detection rate. Nucl. Instrum. Methods Phys. Res., Sect. A **477**, 244–249 (2002)
7. A.E. Cameron, D.F. Eggers, An ion "Velocitron". Rev. Sci. Instrum. **19**, 605–607 (1948)
8. W.C. Wiley, I.H. McLaren, Time-of-flight mass spectrometer with improved resolution. Rev. Sci. Instrum. **26**(12), 1150–1157 (1955)
9. J.H. Posthumus, The dynamics of small molecules in intense laser fields. Rep. Prog. Phys. **67**(5), 623–665 (2004)
10. N. Marable, G. Sanzone, High-resolution time-of-flight mass spectrometry: theory of the impulsed-focused time-of-flight mass spectrometer. Int. J. Mass Spectrom. Ion Phys. **13**, 185–194 (1974)
11. B.A. Mamyrin, D.V. Shmikk, The linear mass reflectron. Sov. Phys. JETP **49**(5), 762–764 (1979)
12. P. Hansch, L.D.V. Woerkom, High-precision intensity-selective observation of multiphoton ionization: a new method of photoelectron spectroscopy. Opt. Lett. **21**(16), 1286–1288 (1996)
13. L. Robson, K.W. Ledingham, P. McKenna, T. McCanny, S. Shimizu, J.M. Yang, C. Wahlstrm, R. Lopez-Martens, K. Varju, P. Johnsson, J. Mauritsson, Volumetric intensity dependence on the formation of molecular and atomic ions within a high intensity laser focus. J. Am. Soc. Mass Spectrom. **16**, 82–89 (2005)
14. M.A. Walker, P. Hansch, L.D. Van Woerkom, Intensity-resolved multiphoton ionization: circumventing spatial averaging. Phys. Rev. A **57**, R701–R704 (1998)
15. W.A. Bryan, S.L. Stebbings, J. McKenna, E.M.L. English, M. Suresh, J. Wood, B. Srigengan, I.C.E. Turcu, J.M. Smith, E.J. Divall, C.J. Hooker, A.J. Langley, J.L. Collier, I.D. Williams, W.R. Newell, Atomic excitation during recollision-free ultrafast multi-electron tunnel ionization. Nat. Phys. **2**(6), 379–383 (2006)
16. J. McKenna, M. Suresh, B. Srigengan, I.D. Williams, W.A. Bryan, E.M.L. English, S.L. Stebbings, W.R. Newell, I.C.E. Turcu, J.M. Smith, E.J. Divall, C.J. Hooker, A.J. Langley, J.L. Collier, Rescattering-enhanced dissociation of a molecular ion. Phys. Rev. A **74**, 043409 (2006)
17. L.J. Frasinski, K. Codling, P.A. Hatherly, Covariance mapping: a correlation method applied to multiphoton multiple ionization. Science **246**, 1029–1031 (1989)
18. K. Codling, L.J. Frasinski, Dissociative ionization of small molecules in intense laser fields. J. Phys. B: At. Mol. Opt. Phys. **26**, 783–809 (1993)
19. L.J. Frasinski, P.A. Hatherly, K. Codling, M. Larsson, A. Persson, C.G. Wahlstrom, Multielectron dissociative ionization of co_2 in intense laser fields. J. Phys. B: At. Mol. Opt. Phys. **27**, L109–L114 (1994)
20. L.J. Frasinski, M. Stankiewicz, P.A. Hatherly, G.M. Cross, K. Codling, A.J. Langley, W. Shaikh, Molecular h_2 in intense laser fields probed by electron-electron, electron-ion, and ion-ion covariance techniques. Phys. Rev. A **46**, R6789–R6792 (1992)
21. J.L. Hansen, J.H. Nielsen, C.B. Madsen, A.T. Lindhardt, M.P. Johansson, T. Skrydstrup, L.B. Madsen, H. Stapelfeldt, Control and femtosecond time-resolved imaging of torsion in a chiral molecule, J. Chem. Phys. **136**, 204310–204310-10 (2012)
22. A.T.J.B. Eppink, D.H. Parker, Velocity map imaging of ions and electrons using electrostatic lenses: application in photoelectron and photofragment ion imaging of molecular oxygen. Rev. Sci. Instrum. **68**(9), 3477 (1997)

23. M.J.J. Vrakking, An iterative procedure for the inversion of two-dimensional ion/photoelectron imaging experiments. Rev. Sci. Instrum. **72**, 4084–4089 (2001)
24. V. Dribinski, A. Ossadtchi, V.A. Mandelshtam, H. Reisler, Reconstruction of abel-transformable images: the gaussian basis-set expansion abel transform method. Rev. Sci. Instrum. **73**, 2634–2642 (2002)
25. G.A. Garcia, L. Nahon, I. Powis, Two-dimensional charged particle image inversion using a polar basis function expansion. Rev. Sci. Instrum. **75**, 4989–4996 (2004)
26. M. Wollenhaupt, M. Krug, J.Köhler, T. Bayer, C. Sarpe-Tudoran, T. Baumert, Three-dimensional tomographic reconstruction ofultrashort free-electron wave packets. Appl. Phys. B **95**, 647–651 (2009)
27. J. Maurer, D. Dimitrovski, L. Christensen, L.B. Madsen, H. Stapelfeldt, Molecular-frame 3D photoelectron momentum distributions by tomographic reconstruction. Phys. Rev. Lett. **109**, 123001 (2012)
28. S. Kauczok, N.G/"adecke, A.I. Chichinin, M. Veckenstedt, C. Maul, K.-H. Gericke, Three-dimensional velocity map imaging: setup and resolution improvement compared to three-dimensional ion imaging. Rev. Sci. Instrum. **80**, 083301–083301-10 (2009)
29. D. Strasser, X. Urbain, H.B. Pedersen, N. Altstein, O. Heber, R. Wester, K.G. Bhushan, D. Zajfman, An innovative approach to multiparticle three-dimensional imaging. Rev. Sci. Instrum. **71**, 3092–3098 (2000)
30. L. Dinu, A.T.J.B. Eppink, F. Rosca-Pruna, H.L. Offerhaus, W.J. van der Zande, M.J.J. Vrakking, Application of a time-resolved event counting technique in velocity map imaging. Rev. Sci. Instrum. **73**, 4206–4213 (2002)
31. J. Eland, Photoelectron-photoion coincidence spectroscopy: I. basic principles and theory. Int. J. Mass Spectrom. Ion Phys. **8**, 143–151 (1972)
32. C. Danby, J. Eland, Photoelectron-photoion coincidence spectroscopy: II. design and performance of a practical instrument. Int. J. Mass Spectrom. Ion Phys. **8**, 153–161 (1972)
33. R.E. Continetti, Coincidence spectroscopy. Annu. Rev. Phys. Chem. **52**(1), 165–192 (2001)
34. J. Ullrich, R. Moshammer, A. Dorn, R. Dörner, L.P.H. Schmidt, H. Schmidt-Böcking, Recoil-ion and electron momentum spectroscopy: reaction-microscopes. Rep. Prog. Phys. **66**, 1463–1545 (2003)
35. J. Ullrich, R. Moshammer, M. Unverzagt, W. Schmidt, P. Jardin, R. Olson, R. Dorner, V. Mergel, H. Schmidt-Bocking, Ionization collision dynamics in 3.6 MeV/u ni24+ on he encounters. Nucl. Instrum. Methods Phys. Res., Sect. B **98**, 375–379 (1995)
36. R. Moshammer, M. Unverzagt, W. Schmitt, J. Ullrich, H. Schmidt-Bocking, A 4π recoil-ion electron momentum analyzer: a high-resolution microscope for the investigation of the dynamics of atomic, molecular and nuclear reactions. Nucl. Instrum. Methods Phys. Res., Sect. B **108**, 425–445 (1996)
37. G. Scoles, D. Bassi, U. Buck, D.C. Laine (eds.), *Atomic and Molecular Beam Methods*, vol. 1. (Oxford University Press, Oxford, 1988)
38. R.D. Zucker, O. Biblarz, *Fundamentals of gas dynamics*, 2nd edn. (Wiley, 2002)
39. D.M. Lubman, C.T. Rettner, R.N. Zare, How isolated are molecules in a molecular beam, J. Phys. Chem. (United States) **86**, 7 (1982)
40. O.F. Hagena, Cluster formation in expanding supersonic jets: effect of pressure, temperature, nozzle size, and test gas. J. Chem. Phys. **56**, 1793 (1972)
41. M. Hillenkamp, S. Keinan, U. Even, Condensation limited cooling in supersonic expansions. J. Chem. Phys. **118**, 8699–8705 (2003)
42. J. Wörmer, V. Guzielski, J. Stapelfeldt, T. Müller, Fluorescence excitation spectroscopy of xenon clusters in the VUV. Chem. Phys. Lett. **159**, 321–326 (1989)
43. R. Campargue, Progress in overexpanded supersonic jets and skimmed molecular beams in free-jet zones of silence. J. Phys. Chem. **88**, 4466–4474 (1984)
44. V. Kumarappan, C.Z. Bisgaard, S.S. Viftrup, L. Holmegaard, H. Stapelfeldt, Role of rotational temperature in adiabatic molecular alignment. J. Chem. Phys. **125**(19), 194309 (2006)
45. A. Miffre, M. Jacquey, M. Buchner, G. Trenec, J. Vigue, Parallel temperatures in supersonic beams: ultra cooling of light atoms seeded in a heavier carrier gas. J. Phys. Chem. **122**, 094308 (2004)

46. A.E. Siegman, *Lasers*, new edn. (University Science Books, London, 1990)
47. M.A. de Araújo, R. Silva, E. de Lima, D.P. Pereira, P.C. de Oliveira, Measurement of gaussian laser beam radius using the knife-edge technique: improvement on data analysis. Appl. Opt. **48**, 393–396 (2009)
48. S.M. Hankin, D.M. Villeneuve, P.B. Corkum, D.M. Rayner, Intense-field laser ionization rates in atoms and molecules. Phys. Rev. A **64**, 013405 (2001)
49. E.A. Gibson, A. Paul, N. Wagner, R. Tobey, S. Backus, I.P. Christov, M.M. Murnane, H.C. Kapteyn, High-order harmonic generation up to 250 ev from highly ionized argon. Phys. Rev. Lett. **92**(3), 033001 (2004)
50. E. Hecht, *Optics*, 4th edn. (Addison-Wesley, 2001)

Chapter 5
Characterisation and Optimisation of Impulsive Molecular Alignment in Mixed Gas Samples

Molecules in gas phase are randomly oriented within the laboratory frame of reference. When interacting with a linearly polarised laser field, the response of the molecular ensemble is thus an average over all orientations of the molecule with respect to the laser polarisation. This makes it impossible to resolve the effect of the molecular structure on the studied laser induced processes. In this chapter laser induced impulsive molecular alignment is presented as a technique to study laser-molecule interactions in the molecular frame. As such, it is one of the corner stones for studying the molecular structure dependence of strong field induced dynamics in CO_2, presented in the following chapters. The aim of the experiments presented in this chapter are two-fold. Firstly, a novel procedure was developed to characterise the quality of molecular alignment in situ in the employed TOF detection scheme. Due to the position-averaging of the anode-based detector, the molecular alignment distribution could not be inferred directly from a TOF measurement. Instead, it had to be extracted by systematically matching numerical simulations to the experimental data. Secondly, this procedure was then used to optimise the alignment quality by seeding the molecular sample in Ar as a carrier gas. It was found that the rotational temperature in the molecular beam depends on the mixing ratio between the two gases. The lowest achievable temperature and thus best molecular alignment was limited by the onset of clustering in the gas jet.

The chapter is organised as follows. Firstly, laser induced molecular alignment is introduced in the context of molecular dynamics research. Secondly the theoretical model that underlies the numerical simulations is presented briefly, which is followed by the experimental approach to detect molecular alignment in a TOF spectrometer. In the remaining sections, the acquired experimental data is presented and analysed by comparing them to the numerical simulations. Here, Sébastien Weber developed the corresponding numerical model and conducted all associated calculations presented in this chapter.

M. Oppermann, *Resolving Strong Field Dynamics in Cation States of CO_2 via Optimised Molecular Alignment*, Springer Theses, DOI: 10.1007/978-3-319-05338-7_5,

5.1 Background

For resolving the molecular structure dependence in dynamical or stereochemical investigations, one has to perform the associated experiments in the molecular frame of reference. Laser induced molecular alignment has become a routine technique for facilitating this, due to the high degrees of achievable alignment, efficient implementation via commercially available pulsed lasers and applicability to a wide range of molecular samples in gas phase [1].

Laser induced alignment of a given molecule is possible if it has an anisotropic polarizability. When interacting with a laser pulse, the resulting anisotropic induced dipole experiences a torque towards the laser pulse polarisation. This aligns the axis of major polarizability of the molecule along the laser pulse polarisation. Quantum mechanically this is described as a coupling between the rotational states of the molecule with the laser field. Through the coupling, the rotational states associated with a free rotor are translated into rotational states that correspond to a rotor librating around the laser polarisation axis. This process can be divided into two regimes: adiabatic and impulsive alignment.

5.1.1 Adiabatic Alignment

In the adiabatic regime, the alignment pulse duration is much longer than the rotational period of the molecule. This implies that at any instant during the interaction the molecule can be approximated to experience a DC field. Upon turn-on of the laser pulse, the DC-field amplitude increases from zero to its maximum. The molecular eigenstate of the field-free Hamiltonian is adiabatically transferred into a so-called *field-induced pendular state* of the complete Hamiltonian, which includes the molecule-laser interaction. These states are aligned around the laser polarisation axis. Upon turn-off of the laser pulse, the molecular ensemble returns to its isotropic distribution.

Qualitatively, the degree of alignment increases with increasing coupling between molecular rotational states and decreasing rotational temperature of the molecular ensemble [1]. The coupling of the rotational states depends on the polarizability of the molecule and on the field amplitude. Less polarisable molecules thus require larger laser intensities to achieve a similar degree of alignment. Yet, the applicable field strength is limited by the onset of ionisation. Fortunately, however, nonresonant ionisation rates typically scale with the polarizability, such that less polarisable molecule can be exposed to higher laser intensities.

Decreasing the rotational temperature effectively reduces the random rotational motion of the free molecular ensemble. In a classical picture this is translated into a narrower cone of liberation of the pendular states around the laser polarisation axis and hence a higher degree of alignment. From a quantum mechanical point of view, the thermal ensemble of the molecules leads to an incoherent superposition of the pendular states. The resulting degree of alignment is thus a statistical average over the thermal ensemble. Decreasing the rotational temperature is associated with a narrower thermal ensemble and thus reduces the incoherent averaging. This has been demonstrated experimentally by using the cooling properties of supersonic expansions [2] and the selection of rotational states via inhomogeneous electric fields [3], for example.

Furthermore, successful 3-dimensional alignment of several large molecules has been reported [4, 5] using elliptically polarised fields. Here, the minor axis of ellipticity aligns the minor axis of polarizability of the molecule. This confines the molecular plane spanned by its axes of polarizability to the laser polarisation plane. Orientation of asymmetric molecules like Iodobenzene has been achieved via static electric fields, interacting with the permanent dipole of the molecule [5].

5.1.2 Impulsive Alignment

In the impulsive regime, the laser pulse duration is much shorter than the characteristic rotational period of the molecule [6]. Here, the laser pulse coherently couples the rotational states of the molecule and thus creates a rotational wavepacket that is initially aligned along the laser pulse polarisation. After the interaction with the laser pulse, the rotational wavepacket then de- and rephases at times characteristic to the molecule. Constructive interference of the rotational states is associated with alignment of the wavepacket along the initial laser pulse polarisation and referred to as an alignment revival. Destructive interference on the other hand leads to the molecular ensemble being confined in a plane perpendicular to the polarisation direction and is referred to as an anti-revival. Full alignment revivals take place every rotational period T_0 of the molecule, with partial revivals every half of this period. Depending on the type of molecule, quarter revivals may be observed as well. Recently, however, even higher order fractional revivals at 1/8th, 1/12th and 1/16th of the rotational period have been observed [7]. The periodic revival structure persists until the coherence of the wavepacket is lost [8, 9]. This typically happens through collisions with other molecules and can be avoided by maintaining a low molecular density in the laser-molecule interaction region.

Like in the case of adiabatic alignment, increasing degrees of impulsive alignment are achieved with increasing molecular polarizability and decreasing rotational temperature of the molecular ensemble [6]. However, in the impulsive regime the degree of rotational excitation solely depends on the total energy deposited in the molecular system. This implies that at a given intensity, the maximum achievable degree of alignment is limited by the pulse length [6]. The increase of the pulse length is limited, however, by the onset of the adiabatic regime. Here, rotational deexcitation takes place in the falling edge of the laser pulse, which leads to a less pronounced revival structure. This is displayed in part (b) of Fig. 5.1.

Several experimental approaches have been designed for improving molecular alignment in the impulsive regime. Ghafur et al. have reduced the initial rotational

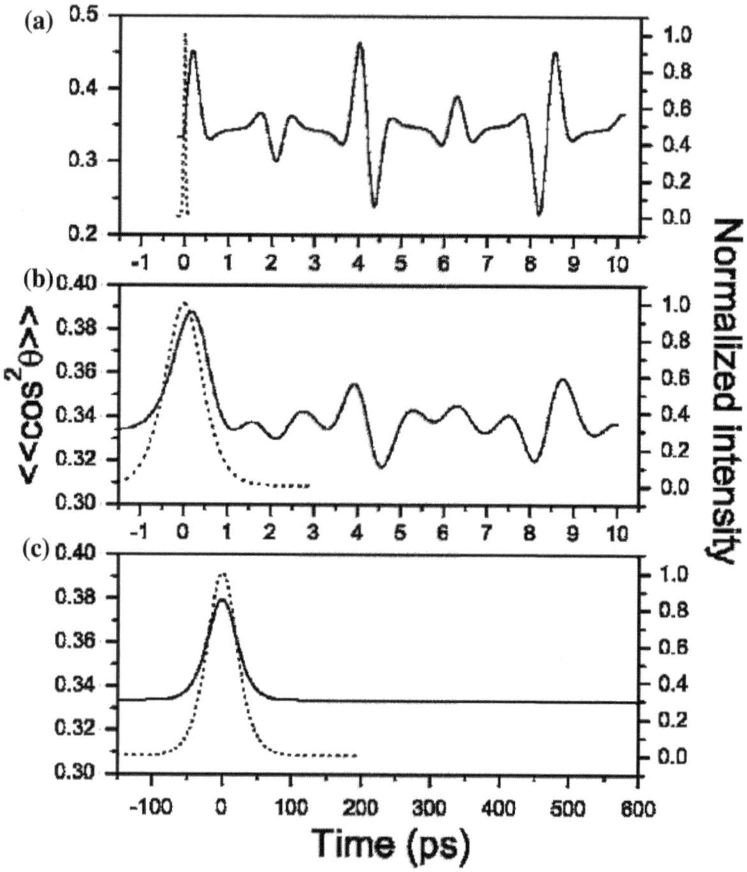

Fig. 5.1 Calculated alignment quality in N_2 at 50 K as a function of time for different alignment pulse durations. The pulse duration τ and peak intensity I_0 used for the simulations were **a** $\tau = 50$ fs, $I_0 = 2.5 \times 10^{13}$ Wcm^{-2}, **b** $\tau = 1$ ps, $I_0 = 2.5 \times 10^{12}$ Wcm^{-2}, **c** $\tau = 50$ ps, $I_0 = 2.5 \times 10^{12}$ Wcm^{-2} (Reprinted with permission from [6]. Copyright (2005) by the American Physical Society)

state distribution via a hexapole state selector and optimised the alignment quality by shaping the alignment pulse [10]. Furthermore, appropriately time-delayed trains of alignment pulses have been employed to enhance the alignment quality [11]. Orientation of the molecule has been achieved by employing two-colour alignment pulses [12] and the possibility of inducing field-free 3-dimensional alignment via two orthogonally polarised pulses has been proposed [13] and demonstrated [14].

5.1.3 Adiabatic Versus Impulsive Alignment

Both laser induced alignment techniques have their advantages and disadvantages. However, one's choice is mostly limited by the target molecule. Impulsive alignment becomes less effective with increasing size of the molecule. This is due to the loss of rotational coherence through rovibronic couplings within the molecule that become more likely with increasing molecular size. Adiabatic alignment on the other hand becomes less effective with decreasing weight of the molecule. This is because a lighter molecule requires a higher alignment pulse intensity for achieving a similar degree of molecular alignment compared to a heavier molecule. Yet, the maximum applicable intensity is limited by the onset of ionisation or dissociation. For this, impulsive alignment is typically used for fixing diatomic and triatomic linear molecules in space. Adiabatic alignment is mostly used for aligning larger molecules like substituted benzenes.

The main advantage of impulsive alignment is that it offers the possibility of probing an aligned molecular sample under field-free conditions. This means that the molecular structure can be resolved without altering the probed molecular system. For this, an alignment pulse creates the rotational wavepacket, which is probed by a second laser pulse at a time delay corresponding to the revival time. Under similar conditions, however, impulsive alignment typically leads to significantly lower degrees of molecular alignment compared to the adiabatic case. It was found that this is due to phase mismatches of the rotational states in the impulsively induced rotational wavepacket [6]. However, due to the shorter pulse duration, higher intensities can typically be applied to the molecule before reaching the ionisation or dissociation threshold, compared to the long pulses used in adiabatic alignment. This implies that in practice, impulsive alignment may reach similar degrees of alignment compared to the adiabatic case.

For this thesis, the molecular structure dependence of strong field processes was studied in CO_2. Because of the target molecule, impulsive alignment was the more suitable choice, with the additional advantage of field-free alignment. The remainder of this chapter will thus only be concerned with impulsive alignment of small linear molecules.

5.1.4 Detecting and Characterising Molecular Alignment

Quantitatively, the degree of molecular alignment is usually characterised via the expectation value $\langle\cos^2\theta\rangle$ of the alignment distribution, where θ is the angle between the laser polarisation and the molecular axis. In this case, $\langle\cos^2\theta\rangle = 1$ means perfect alignment, $\langle\cos^2\theta\rangle = 0$ perfect anti-alignment and $\langle\cos^2\theta\rangle = 1/3$ random alignment. For measuring $\langle\cos^2\theta\rangle$ directly, one has to record the bond axis distribution of the molecular ensemble. This can be achieved by probing the aligned molecular ensemble via a second high intensity laser pulse inducing Coulomb explosion and recording the angular distribution of the resulting fragments via a VMI spectrometer (see for example [2, 10, 15]). As was pointed out in Sect. 2.2.4, in the Coulomb explosion regime the fragments are ejected along the molecular axis. For this, the angular confinement of the fragment ions at the detector is directly related to the bond axis distribution.

Nevertheless, molecular alignment can be identified qualitatively in any process that is sensitive to the orientation of the molecular structure in the laboratory frame. This has been demonstrated for strong field ionisation in form of high harmonic generation [16, 17] and Kerr effect birefringence induced processes [18], where the recorded signal depends on the alignment of the molecule with respect to the linear laser polarisation. Revival structures have further been obtained from Coulomb explosion detected via a TOF spectrometer [19], where the detected fragmentation yields depend on the orientation of the molecule with respect to the spectrometer axis. The latter technique was employed in the experiments presented in this chapter and is discussed in detail in Sect. 5.4. As these techniques are sensitive to the molecular alignment distribution, the time evolution of the rotational wavepacket can be recorded qualitatively. This means that the revival structure is obtained, yet its scaling in terms of the degree of molecular alignment $\langle\cos^2\theta\rangle$ cannot be extracted directly from the data.

For the experiments presented in this thesis, a systematic procedure was developed to characterise the molecular alignment via a TOF spectrometer. The scaling of the revival structure in terms of the degree of molecular alignment was obtained by carefully matching numerical simulations to the experimental data. The main advantage of this approach is that the molecular alignment distribution can be determined in situ without the need of an additional and possibly more complex detection system. Furthermore, if molecular alignment is used in an experiment, the developed procedure allows for its full characterisation under the employed experimental conditions. This also enables one to systematically optimise the alignment quality by varying the alignment pulse parameters and rotational cooling of the gas sample. For the experiments presented in this thesis it was not possible to alter the alignment pulse parameters. In particular, one would stretch the alignment pulse duration via a dispersive element and increase the pulse energy to obtain a sufficient degree of rotational state coupling. Especially when employing 800 nm probe pulses, however, the alignment pulse energy was limited by the available output energy from the CPA laser system. This made it impossible to further stretch the alignment pulse duration. For this, the optimisation of the molecular alignment was concentrated on the rotational cooling of the sample gas. In particular, the sample molecule was seeded in Ar as a carrier gas and the mixing ratio optimised for the best rotational cooling in the supersonic gas expansion. The analysis and results of this optimisation are presented in this chapter.

5.2 Theory of Impulsive Alignment

In this section the most important aspects of the theory for describing impulsive molecular alignment are presented. It is based on the properties of a quantum mechanical rigid rotator in a laser field (see for example [20, 21]) and was applied in the late 1990s to the case of impulsive alignment [8, 9]. The discussion of the theory presented in this chapter mostly follows [22].

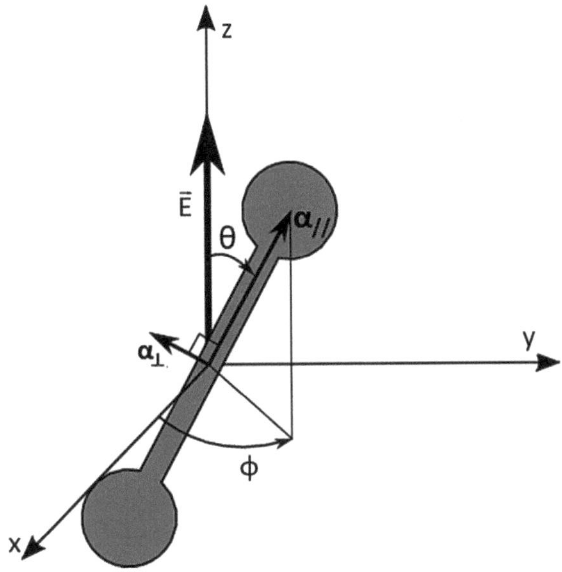

Fig. 5.2 Coordinate system for describing the interaction between a laser field **E** polarised along the z-axis and a linear molecule. θ is the angle of the molecular axis with respect to the electric field vector and $\alpha_{//}$ and $\alpha_{\perp}$ refer to the polarizability parallel and perpendicular to the molecular axis, respectively

5.2.1 Theoretical Model

Similar to the discussion in Sect. 2.2.3, the time evolution of a molecule in a laser field is described by its total wavefunction $\Psi(t)$, which fulfills the time dependent Schrödinger equation (TDSE) in atomic units:

$$i\dot{\Psi}(t) = H_{tot}(t)\Psi(t) \tag{5.1}$$

with $H_{tot}(t) = H_e + H_{vib} + H_{rot} + H_L - \boldsymbol{\mu} \cdot \mathbf{E}(t)$. Here, the first three terms in the total Hamiltonian refer to the electronic, vibrational and rotational degrees of freedom, respectively. H_L refers to the laser field and the last term describes the laser-molecule interaction within the dipole approximation. For the treatment of impulsive molecular alignment, one further assumes that the laser field is non-resonant with electronic or vibrational transitions and can be treated perturbatively. This implies that H_e, H_{vib} and H_L can be neglected. The relevant Hamiltonian thus consists of the rotational kinetic energy of the molecule and its potential energy caused by torque imposed by the laser field.

Figure 5.2 displays the geometry of a linear molecule with anisotropic polarizability. The electric field vector **E** is aligned along the z-coordinate axis and θ is the angle between the laser polarisation direction and the molecular axis.

Rotational Hamiltonian

With B_0 the rotational constant of the molecule and $\mathbf{J}$ the angular momentum operator, the rotational Hamiltonian is written as (see for example [21]):

$$H_{rot} = B_0 \mathbf{J}^2 \tag{5.2}$$

The resulting associated rotational eigenstates $|J, M\rangle$ form the angular momentum basis with the quantum number $J \geq 0$ referring to the total orbital angular momentum and $|M| \leq J$ referring to its projection onto the coordinate z-axis. In space coordinates, the rotational eigenstates correspond to the spherical harmonics $\Upsilon_J^M(\theta, \phi)$. The associated eigenenergies E_J are given by:

$$E_J = B_0 J (J+1) - D_0 J^2 (J+1)^2 \tag{5.3}$$

Here, the first term is the solution obtained from the unperturbed time independent Schrödinger equation with H_{rot} assuming a fully rigid rotator. The second term accounts for rovibrational couplings that are associated with a stretching of the molecular bond at high angular momentum [20]. For this D_0 is the centrifugal constant that depends on the strength of the bond of the molecule.

Potential Energy Term

The polarizability tensor $\boldsymbol{\alpha}$ of a molecule measures the charge distribution distortion due to the interaction with an electric field $\mathbf{E}(t)$ and hence the total dipole moment contains a laser induced term (see for example [23]):

$$\boldsymbol{\mu} \approx \boldsymbol{\mu}_0 + \frac{1}{2}\boldsymbol{\alpha}\mathbf{E}(t) \tag{5.4}$$

with μ_0 the permanent dipole of the molecule. Note that the equation is an approximation and in principle the induced dipole moment may also contain nonlinear terms. However, for the typical intensities employed for impulsive alignment, neglecting higher order terms is appropriate. For the following discussion, the permanent dipole is neglected. This is because the associated potential energy follows the field instantaneously on the time scale of a molecular rotation. When averaging over the laser field oscillations, the potential energy changes cancel out. Referring to the field induced potential energy as $U(\theta, t)$, one obtains:

$$U(\theta, t) = -\boldsymbol{\mu} \cdot \mathbf{E}(t) \tag{5.5}$$

$$= -\frac{1}{4}|E|^2 f(t)^2 \left(\Delta\alpha \cos^2\theta + \alpha_\perp\right) \tag{5.6}$$

with $\Delta\alpha = \alpha_\parallel - \alpha_\perp$, E the amplitude and $f(t)$ the envelope of the laser pulse. Briefly, for arriving at the final expression, one would express the polarizability tensor in the molecular frame where it is diagonal. Choosing the internuclear axis to be the z-axis and due to the molecule's cylindrical symmetry, one can set $\alpha_{xx} = \alpha_{yy} \equiv \alpha_\perp$ and $\alpha_{zz} \equiv \alpha_\parallel$. For the same reasons it can also be inferred that after transforming the diagonal polarisation tensor back to the laboratory frame, it will only depend on the polar angle θ out of the three Euler angles of the molecule.

Rotational Coupling and Resulting Wavepacket

Following the above discussion, the total Hamiltonian for describing the rigid rotor in a laser field becomes:

$$H_{tot} = B_0\mathbf{J}^2 + U(\theta, t) \tag{5.7}$$

Here, the potential energy term is responsible for the coupling of different rotational states. The allowed couplings are then obtained by evaluating whether the respective transition matrix element $C_{J,J',M,M'}$ in the angular momentum basis is nonzero:

$$C_{J,J',M,M'} = \langle J', M'|U(\theta, t)|J, M\rangle \propto \langle J', M'|\cos^2\theta|J, M\rangle \tag{5.8}$$

As $\cos^2\theta$ is an even function in coordinate space, $C_{J,J',M,M'}$ is only nonzero if the coupled states have the same parity. This implies that even and odd states are not coupled and $\Delta J = J' - J = 0, \pm 2$. Such transitions are commonly referred to as being of the Raman type. Due to the cylindrical symmetry of the rigid rotor, $C_{J,J',M,M'}$ is independent of ϕ and states with different magnetic quantum numbers cannot be coupled. This implies $\Delta M = M' - M = 0$.

The final solution $|\Psi(t)\rangle$ to the TDSE associated with the total Hamiltonian from Eq. 5.7 is a superposition of the rotational states coupled by the laser pulse, with eigenenergies given by Eq. 5.3. It is expanded in the angular momentum basis:

$$|\Psi(t)\rangle = \sum_J a_{JM}(t)e^{-iE_J/\hbar t}|J, M\rangle \tag{5.9}$$

As was pointed out earlier, the wavepacket de- and rephases periodically until the rotational coherence is lost. The molecular distribution at a time $t = t_0$ after the alignment pulse can be characterised by the expectation value $\langle\cos^2\theta\rangle = \langle\Psi(t_0)|\cos^2\theta|\Psi(t_0)\rangle$ with the initial condition that the molecule is in a given rotational state $|J, M\rangle$ previous to the alignment pulse interaction. Each initial condition thus leads to a coupling between states $|J, M\rangle$ and $|J + 2, M\rangle$. The temporal evolution of this superposition state is thus determined by the beating frequency ω_J of the coupled states:

$$\omega_J = E_{J+2} - E_J = B_0(4J + 6) \tag{5.10}$$

The temporal evolution of the rotational wavepacket is thus determined by the spectrum of beating frequencies resulting from the initial distribution of rotational states (as initial conditions) and Eq. 5.10. Full wavepacket revivals take place at each rotational period:

$$T_{rot} = (2B_0 c)^{-1} \tag{5.11}$$

where c is the vacuum speed of light.

5.2.2 *Thermal Ensemble*

The initial distribution of rotational states can be modeled by a Boltzmann distribution with the molecular gas represented by a thermal ensemble at rotational temperature T. The initial population of a particular state $|J, M\rangle$ is thus given by a probability distribution P_J:

$$P_J = \frac{1}{Z}(2J+1)\, g_J e^{-\frac{E_J}{k_B T}} \tag{5.12}$$

where Z is a normalisation factor and k_B the Boltzman constant. The factor $(2J+1)$ accounts for the $(2J+1)$-fold degeneracy of a rotational state with $J > 1$ if no external field is applied. For the state distribution this means that such a state occurs $(2J+1)$ times more likely than a state with $J = 0$. The factor g_J arises from the nuclear spin statistics of symmetric linear molecules. It determines the distribution of even and odd J states within the ensemble and is essentially based on the Pauli exclusion principle [20].

The rotational wavepacket evolution of the thermal ensemble is thus a statistical average over the initial conditions given by the initial state distribution. The alignment distribution is therefore described by a statistical average of the expectation value $\langle \cos^2\theta \rangle$ over the thermal ensemble:

$$\left\langle\left\langle \cos^2\theta \right\rangle\right\rangle_{stat} = \frac{1}{Z}\sum_{J,M} g_J e^{-\frac{E_J}{k_B T}} \left\langle \cos^2\theta \right\rangle \tag{5.13}$$

For convenience, $\langle\langle \cos^2\theta \rangle\rangle_{stat}$ will be referred to as $\langle \cos^2\theta \rangle$ for the remainder of this chapter. In the statistical average, the superposition of the individual revival structure for each initial condition is incoherent and thus reduces the overall degree of molecular alignment. This implies that reducing the number of states in the initial state distribution improves the degree of molecular alignment, due to the reduced number of incoherent superpositions. This effect will be examined in more detail via numerical simulations in Sect. 5.3.

Table 5.1 Values for the molecular properties used in the numerical simulations (data taken from http://webbook.nist.gov)

Molecule	B_0 (cm^{-1})	D_0 (10^{-6} cm^{-1})	$\Delta\alpha$ (Å^3)	g_{even}	g_{odd}
N_2	1.9896	5.76	1.0	2	1
CO_2	0.3902	0.135	2.0	1	0
O_2	1.4297	4.839	1.15	0	1

5.2.3 Numerical Simulations

Simulations were performed for three different molecules; N_2, CO_2 and O_2. The values for the required molecular properties are listed in Table 5.1. The temporal evolution of the rotational wavepacket in form of the time-dependent expectation value $\langle\cos^2\theta\rangle$ was obtained by numerically integrating the TDSE. For this, a Runge-Kutta algorithm was employed. The shape of the laser pulse was chosen to be:

$$\mathcal{E}(t) = \mathcal{E}\sin\left(\frac{\pi t}{2\delta t}\right) \tag{5.14}$$

where δt denotes the intensity FWHM duration of the laser pulse. Even though a Gaussian envelope would be a more appropriate choice for describing the experimental pulse shape, it was found that both pulse forms result in very similar results in terms of the rotational wavepacket dynamics. However, the shape employed in Eq. 5.14 had the advantage of exactly fixing the laser-molecule interaction time within the simulation and thus enabled a definite starting time for the field-free evolution of the wavepacket. Furthermore, the pulse shape allowed for an exact solution of the temporal integration involved in solving the TDSE. This made the numerical calculations more efficient. The temporal wavepacket evolution was then Fourier transformed for obtaining the relative distribution of beating frequencies of the involved rotational states.

5.3 Impact of the Rotational Temperature

It was pointed out previously that the quality of molecular alignment improves with decreasing rotational temperature of the molecular ensemble. As was discussed in Sect. 4.4.4, seeding of the molecule in an atomic carrier gas may lead to enhanced rotational cooling of the seed during a supersonic expansion. Yet, the cooling efficiency sensitively depends on the atomic carrier species [24] and the properties of the gas jet like nozzle dimensions, molecular beam skimming [25, 26] and backing pressure [2, 24]. Typically however, the geometrical characteristics of the gas jet such as nozzle and skimmer properties and distances cannot be varied. Additionally, increasing the backing pressure is usually severely limited by the pumping speed

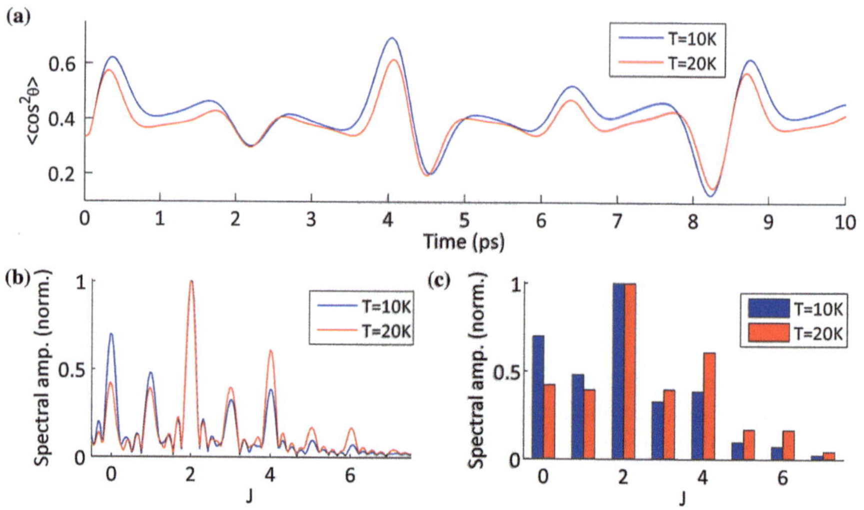

Fig. 5.3 **a** Temporal evolution of rotational wavepackets in N_2 induced by a 70 fs pulse at 3×10^{13} Wcm^{-2}. The calculation was performed for two rotational temperatures at 10 and 20 K. **b** Displays the associated spectrum of rotational states obtained via Fourier transforming the revival structure. **c** Shows the same spectrum as a histogram

of the vacuum system and the valve type. The gas jet used for this thesis employed a piezo valve, which limited the applicable backing pressure to 3 bar. The above limitations imply, however, that seeding the molecule in an atomic carrier gas is an efficient method for decreasing the rotational temperature of the molecular ensemble, because the gas jet setup does not need to be changed. Nevertheless, it has already been discussed that the achievable degree of cooling is limited by the onset of clustering in the gas expansion. Given that the operation of the gas valve was excluded for controlling the cooling mechanism, one has to carefully adjust the mixing ratio of the seeded gas mix for achieving the lowest rotational temperature whilst avoiding clustering. Changing the mixing ratio of the gases in the mix changes their partial pressures for the supersonic expansion. Decreasing the partial pressure of the seed thus increases the rotational cooling, as the carrier gas reaches lower temperatures due to its increased partial pressure.

In order to be able to systematically investigate the alignment quality as a function of rotational temperature, numerical simulations were conducted. Figure 5.3a compares the temporal evolution of two rotational wavepackets at rotational temperatures of 10 and 20 K, respectively. They were induced in N_2 via a 70 fs laser pulse at an intensity of 3×10^{13} Wcm^{-2}. As expected, the simulations show that decreasing the rotational temperature leads to a significant improvement of the degree of alignment at the wavepacket revivals.

However, this effect is not easily quantified in an experiment where the revival structure is only measured qualitatively. This is because the scaling of the detected signal with $\langle \cos^2 \theta \rangle$ might be too weak to unambiguously compare revival structures for determining differences in the alignment quality and thus rotational temperature.

Yet, the temporal evolution of the wavepacket contains a discrete spectrum of beating frequencies that is directly related to the initial rotational state distribution. For comparing the spectra corresponding to each rotational temperature, the Fourier transform (FT) of the wavepacket evolution was normalised with respect to the highest populated state after subtracting any possible DC offset. The frequency axis is scaled in units of the rotational quantum number, such that:

$$J = \frac{\omega_J}{4B_0} - \frac{3}{2} \tag{5.15}$$

The resulting spectra are displayed in Fig. 5.3b. For a clearer presentation, the spectrum is plotted as a histogram, where only the maximum value of each peak is used. This is displayed in Fig. 5.3c and is the preferred representation for the rest of this chapter. Clearly, lowering the rotational temperature implies that the rotational state distribution is shifted towards lower states. Due to the discreteness of the spectrum, comparing experimental data in this representation is less ambiguous. Because of the linearity of the FT one always obtains the same relative rotational state distribution, even if the associated temporal wavepacket evolution is linearly scaled. This implies that the analysis of experimental data obtained from qualitative alignment detection techniques is more appropriate in the frequency domain. This has been used in Sect. 5.6 to accurately match theoretical calculations to experimental data for extracting rotational temperature and the alignment distribution from experiments.

5.4 Experiment

Pump-probe experiments were conducted for recording the temporal evolution of rotational wavepackets in the molecules N_2, CO_2 and O_2. For this, the pump or alignment pulse induced a rotational wavepacket. At a given time delay, a high intensity probe pulse induced Coulomb explosion of the molecular ensemble. The resulting molecular fragments were detected via a TOF spectrometer. By scanning the pump-probe delay, the revival structure of the rotational wavepacket could be recorded. This was done for different gas mixing ratios, with the molecular sample seeded in Ar. The data was Fourier transformed for comparing the rotational state distributions as a function of the mixing ratio. In this section, the employed experimental setup and data analysis procedure are presented.

5.4.1 Setup and Procedure

The setup consisted of three parts: the pump-probe beamline, the molecular beam and gas mix delivery and the TOF spectrometer. It is schematically displayed in Fig. 5.4a. The beamline employed for the experiment used 800 nm probe and 800 nm

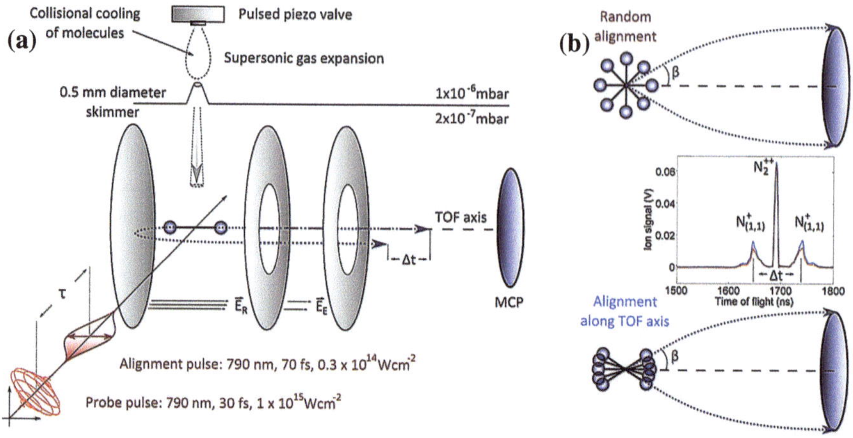

Fig. 5.4 **a** Schematic illustration of the TOF detection setup including the two delayed laser pulse for inducing and probing the rotational wavepacket evolution. **b** Illustration of the detection principle for identifying molecular alignment in TOF spectrometer, which makes use of the finite acceptance angle of the detector for fragments with kinetic energy release

alignment pulses and was presented in Sect. 4.3.2. The probe pulse had a duration of 30 fs and an intensity of 1×10^{15} Wcm^{-2} at the interaction region. The alignment pulse was 70 fs long with an intensity of 3×10^{13} Wcm^{-2}. In both cases the intensity was estimated via the measured pulse length, energy and focal spot diameters. The alignment pulse intensity was controlled by clipping its beam via an iris prior to focussing. The iris was adjusted to keep the ionisation signal from the alignment pulse alone more than an order of magnitude lower than the signal from the probe. The alignment pulse was linearly polarised along the spectrometer axis. This did not require a half-wave plate in the alignment arm. The probe pulse was circularly polarised via a quarter-wave plate in the probe arm. This was necessary to avoid additional dynamic alignment induced by the probe pulse.

The molecular beam was operated at a repetition rate of 47 Hz and an effective valve opening time of 30 μs. Throughout all measurements, the backing pressure was kept at 2.1 ± 0.2 bar and the opening distance of the valve was set to maintain a background pressure in the source chamber of 1×10^{-6} mbar. This resulted in a background pressure in the interaction chamber below 2×10^{-7} mbar during operation of the gas jet. The preparation of a mixed gas sample was presented in Sect. 4.3.4. For each investigated molecule gas mixes were produced with the concentration of the molecule R_{mol} given by the partial pressures of each component $R_{mol} = p_{mol}/p_{Ar}$. For N_2 the pure molecular sample ($R_{N_2} = 1$) was compared to four mixes ($R_{N_2} = 0.3, 0.2, 0.1, 0.02$). In the case of CO_2 and O_2 the pure sample was only compared to a 10 % mix. The reproducibility of the mixed samples was tested by comparing the experimental data obtained from two independently produced mixes of the same ratio. The experimental results did not show a difference.

Table 5.2 Details of the pump-probe delay scans performed for recording the revival structure of rotational wavepackets induced in N_2, CO_2 and O_2

Molecule	T_{rot}	τ_i (ps)	τ_f (ps)	$\Delta\tau$ (ps)	Shots
N_2	8.38	−0.5	20	0.1	1,000
CO_2	42.74	−0.6	90	0.2	800
O_2	11.67	−0.5	27	0.1	800

The TOF spectrometer was operated under Wiley McLaren space focussing conditions as presented in Sect. 4.2. For recording the time evolution of the rotational wavepacket in a sample molecule, TOF spectra were recorded at different pump-probe delays. Length, step size and number of shots per data point for a delay scan depended on the sample molecule. They are listed in Table 5.2. In every case, the delay scan started some time before the arrival of the alignment pulse and was done to ensure the recording of the starting time of the free evolution of the induced wavepacket. The scan then covered a delay of at least two rotational periods of the sample molecule. As the total duration of the recorded temporal evolution is directly related to the frequency resolution of its FT, it was determined that this scan length lead to rotational state spectra of sufficient quality. The time step between two consecutive delays was then chosen to be small enough to enable the resolution of the relevant rotational states in the FT of the wavepacket evolution. For this one has to note that resolving higher frequencies in the FT and thus higher rotational states requires decreasing the time step.

TOF spectra were acquired in averaging mode, where the TOF spectrum was averaged over a given number of laser shots by the oscilloscope. The resulting spectrum was then transferred to a computer. The number of shots were chosen to be sufficiently high for achieving an adequate ratio of signal to noise. This was limited by total duration of a single delay scan, which was kept below two hours. This was done to avoid drifts of experimental parameters such as laser intensity and backing pressure during a scan.

The acquired TOF spectra were Fourier filtered to remove high frequency noise. Yields of individual ion species were then obtained by numerical integration of the relevant time window in the TOF spectrum.

5.4.2 Revival Identification

The identification of molecular alignment via a TOF spectrometer relies on its ion momentum sensitivity and resulting finite acceptance angle for fragments ejected with momentum perpendicular to the TOF axis. These aspects were discussed in detail in Sect. 4.1.2. Firstly, in the Coulomb explosion regime, fragmentation channels with high kinetic energy release (KER) are accessed. For a molecule aligned along the TOF axis this splits the corresponding fragment ion peaks in time. This

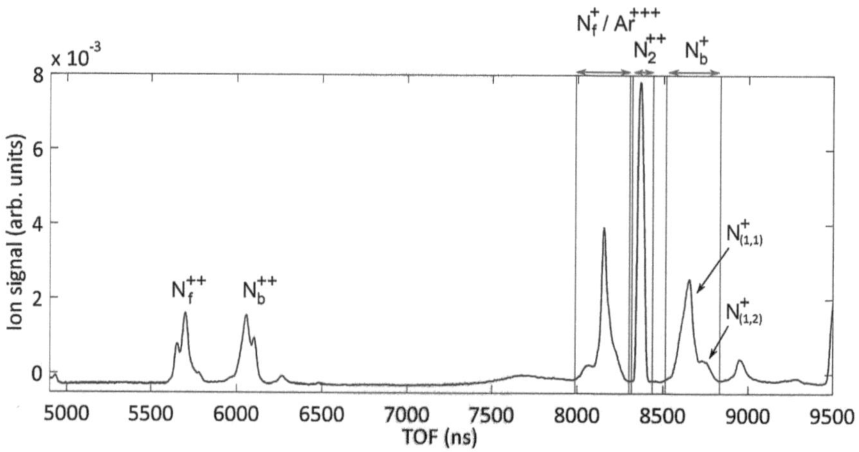

Fig. 5.5 The *upper panel* displays a typical TOF spectrum from a delay scan in N_2 seeded in Ar. The pump-probe delay was 4 ps. The time windows for the numerical integration of the ion peaks used for the data analysis are indicated via *vertical solid lines*

turn-around time Δt is illustrated in part (a) of Fig. 5.4. The TOF spectra in part (b) of the same figure shows the resulting peak splitting for the N^+ fragments into forwards (N_f^+) and backwards (N_f^+) ejected fragment ions. Note that the N_2^{++} peak is not split due to the lack of KER. However, if the molecule is not aligned parallel to the TOF axis, the fragments will possess a momentum component perpendicular to the TOF axis. Due to the finite size of the detector this implies that for a given fragmentation channel, there is a critical alignment angle β, such that only fragments from molecules aligned within this angle around this TOF axis are detected. This enables the identification of molecular alignment and is illustrated in part (b) of Fig. 5.4. With the alignment pulse polarisation parallel to the TOF axis, an alignment revival leads to an alignment distribution confined around the TOF axis. This increases the number of molecules aligned within the critical angle β and thus enhances the ion yield of the corresponding fragment peak.

Figure 5.5 shows a section of a typical TOF spectrum from a revival scan performed for N_2 seeded in Ar. The fragment peaks used for extracting the revival structure correspond to N^+ ions. Due to the high probe intensity, Ar^{+++} ions were formed, which had the same TOF as the N_f^+ ions. Due to the overlapping peaks, only the N_b^+ is used for the analysis. In the figure, the time windows used for numerically integrating the ion yields of the relevant peaks are denoted by red solid lines.

The N^+ yield consisted of two channels, denoted $N^+_{(1,1)}$ and $N^+_{(1,2)}$ Here, the subscript denotes the distribution of the available charge of the parent ion over the two fragments. The $N^+_{(1,2)}$ channel has a higher KER due to the higher Coulombic repulsion of the corresponding fragments and is thus associated with the shoulder of the main peak. The main peak is then associated with the $N^+_{(1,1)}$ channel.

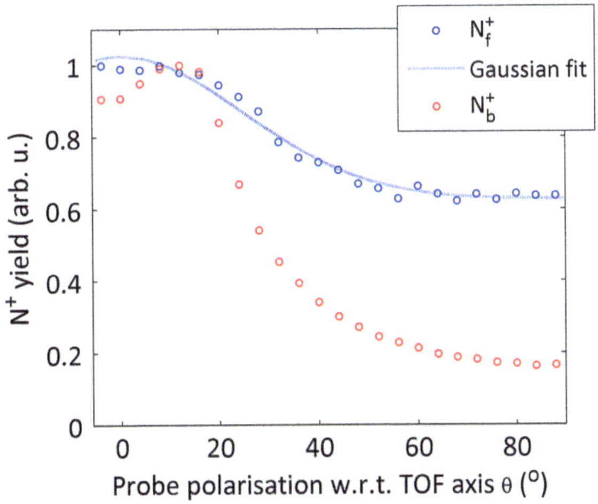

Fig. 5.6 N$^+$ yield obtained from a linearly polarised probe only and plotted as function of angle θ between the polarisation and the TOF axis. The reduction of the detected fragment yield for increasing θ is due to the finite acceptance angle of the TOF spectrometer for fragments ejected away from the TOF axis

The acceptance angle β for the N$^+$ fragments was determined as follows. The alignment pulse was blocked and the quarter-wave plate in the probe arm replaced by a half-wave plate. Due to the high intensity of the probe pulse, the exploded molecules were instantaneously aligned along the now linear probe polarisation. This implied that the resulting fragments were preferably ejected along the probe pulse polarisation. Recording the resulting N$^+$ yield as a function of polarisation angle θ with respect to the TOF spectrometer axis thus reveals the effect of the finite acceptance angle of the TOF spectrometer. The resulting curves for N_f^+ and N_b^+ are plotted in Fig. 5.6.

As the N_f^+ peak overlaps with the Ar^{+++} peak, the variation of the ion yield as a function of θ has a large constant y-offset. However, since Ar^{+++} is not expected to change as a function of polarisation, the decrease in signal is only due to the reduced detection of N_f^+ ions with increasing θ. For this, a Gaussian was fitted to the curve, resulting in a FWHM of 59°. This implies that the acceptance angle is on the order of $\beta \approx 30°$.

The N_b^+ peak behaves as expected as the signal almost vanishes for large polarisation angles θ. However, the signal is significantly reduced for $\theta < 16°$. This is most likely due to N_b^+ ions hitting the repeller plate and thus being scattered too far away from the interaction region to be subsequently detected.

The revival structure of the rotational wavepacket could then be extracted by recording the N_b^+ ion yield as a function of pump-probe delay. In order to account for intensity fluctuations between data points, the N_b^+ ion yield was divided by the N_2^{++} yield. As the N$^+$ fragment yield mostly results from the $N^+_{(1,1)}$ channel, one

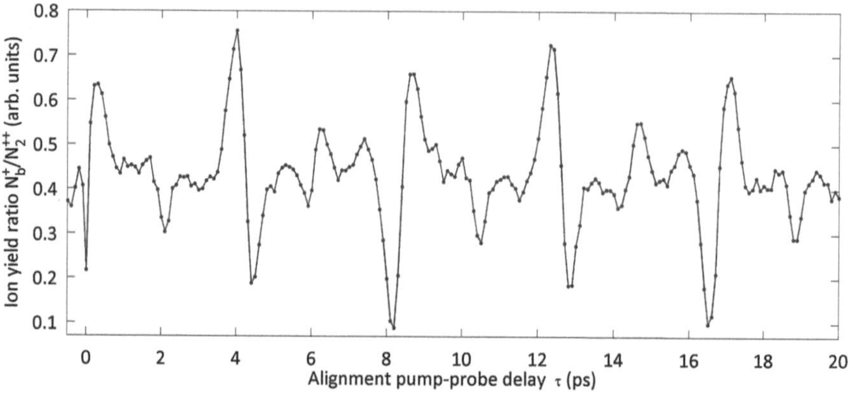

Fig. 5.7 The plot displays the ion yield ratio N_b^+/N_2^{++} as a function of pump-probe delay τ and reveals the revival structure of the induced rotational wavepacket

may assume that the intensity dependence of the N_2^{++} and N^+ yield are very similar. Furthermore, the N_2^{++} yield does not depend on the alignment distribution due to its lack of KER. The resulting ion yield ratio is plotted as a function of pump-probe delay in the lower panel of Fig. 5.7 and clearly reveals the revival structure of N_2. In the case of CO_2 the O^+/CO_2^{++} ratio was used, whilst for O_2 the O^+/O_2^{++} ratio was employed for recording the wavepacket evolution.

5.5 Results

As was pointed in Sect. 5.3, the impact of the rotational temperature onto the alignment quality is visualised best when comparing the rotational state spectra associated with the temporal evolution of the rotational wavepacket. Here, this is done with the experimental data obtained for the molecules N_2, CO_2 and O_2 as a function of mixing ratio R_{mol} with Ar as a carrier gas.

5.5.1 *N*$_2$

Figure 5.8 shows the spectral amplitude distribution for pure N_2 compared to four different gas mixes. One can observe that the distribution is shifted towards lower rotational states for decreasing R_{N_2}. This effect is most strongly pronounced at $J = 0$, where the amplitude increases monotonically for decreasing concentration of N_2 in the gas mix. When examining the behaviour of the spectral amplitudes corresonding to higher coupled rotational states ($J > 3$), one observes a significantly lower population for the gas mixes $R_{N_2} = 0.3, 0.2, 0.1$ compared to pure

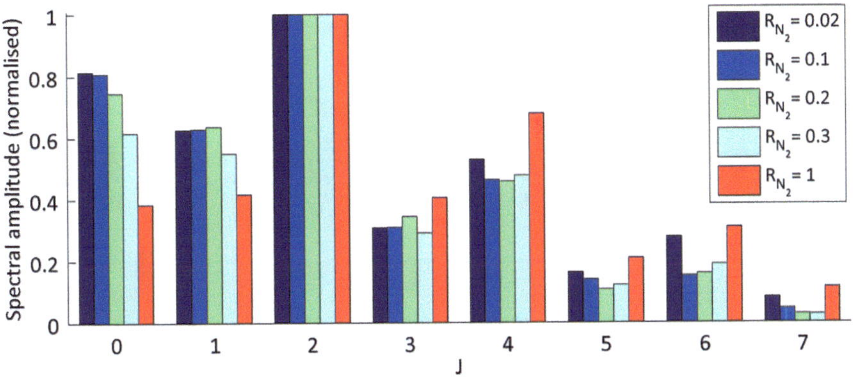

Fig. 5.8 Spectral amplitudes corresponding to the beating frequencies encoded in the temporal evolution of a rotational wavepacket in N_2. The spectra are recorded as a function of concentration of N_2 in a gas mix with Ar. The amplitudes are normalised to the highest peak within each individual spectrum and the frequency axis scaled in units of the rotational quantum number J

N_2. However, the opposite behaviour is observed when decreasing the mixing ratio further to $R_{N_2} = 0.02$. Here, a strong increase of the spectral amplitude for $J > 3$ is displayed. This is particularly visible for the peak at $J = 6$.

5.5.2 *CO_2 and O_2*

Figure 5.9 shows the spectral amplitude distributions measured for pure CO_2 and O_2 compared to the corresponding gas mix with 10 % concentration in Ar. In both cases one observes a clear shift of the distribution towards lower coupled rotational states in accordance with the observations made in N_2. The effect is especially pronounced in CO_2, where the peaks for $J > 10$ almost completely vanish when seeding the molecule in Ar.

5.5.3 *Discussion*

For all investigated molecules, the spectral amplitude distribution is shifted towards lower rotational states when seeding the molecule in Ar, compared to a pure sample. This can be interpreted qualitatively as a decrease in rotational temperature when mixing the molecule in the atomic carrier gas. This is due to the enhanced rotational cooling during the supersonic expansion. In N_2 this effect was investigated systematically by decreasing the concentration of the molecular gas from $R_{N_2} = 0.3$ to $R_{N_2} = 0.02$. It was found that the rotational temperature monotonically decreased up to $R_{N_2} = 0.1$. When decreasing the concentration further for $R_{N_2} = 0.02$, the

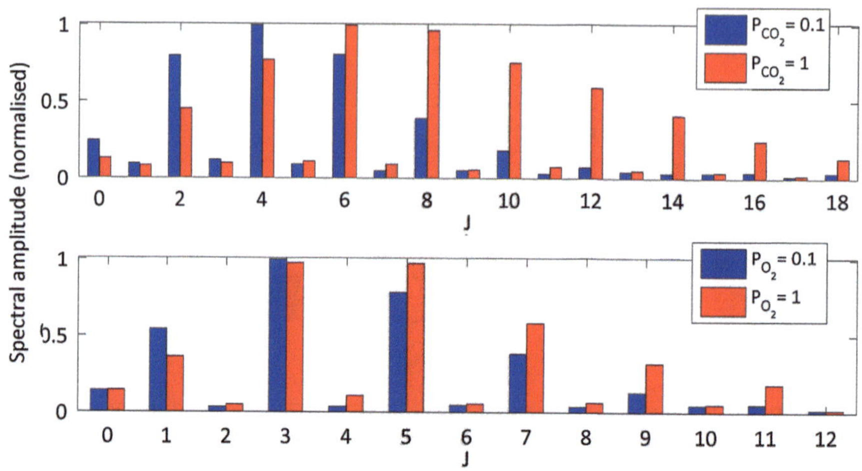

Fig. 5.9 Spectral amplitudes corresponding to the beating frequencies encoded in the temporal evolution of a rotational wavepacket in CO_2 (*upper panel*) and O_2 (*lower panel*). The spectra are recorded as a function of concentration of the respective molecule in a gas mix with Ar. The amplitudes are normalised to the highest peak within each individual spectrum and the frequency axis scaled in units of the rotational quantum number J

high lying spectral amplitudes increased strongly. This must be due either to an increase in rotational temperature or stronger coupling of the rotational states. The latter can only be caused by an increase of alignment pulse intensity. However, this parameter was kept constant throughout the experiment, such that this cause can be excluded. One must therefore conclude that the rotational cooling is optimised for $R_{N_2} = 0.1$. The change in spectral amplitude behaviour for $R_{N_2} = 0.02$ is most probably due to the onset of clustering in the gas jet, as a decrease in concentration of the molecular sample corresponds to an increase of partial pressure of the carrier gas. For $R_{N_2} = 0.02$ one may therefore expect that a critical partial pressure for Ar was reached such that it begins to condensate. In this case, the rigid rotor model is not appropriate anymore for describing the rotational wavepacket, as it is induced in a molecular ensemble that is likely to contain clusters. For this reason, the data for $R_{N_2} = 0.02$ is excluded in the following quantitative analysis.

5.6 Matching Theory to Experiment

In order to extract the alignment distribution and rotational temperature from the experiment, numerical simulations were carefully matched to the experimental data. It was already pointed out that this can be done most efficiently in the Fourier domain. The reason for this is that the same relative spectral amplitude distribution is obtained, even if the associated temporal wavepacket evolution is linearly

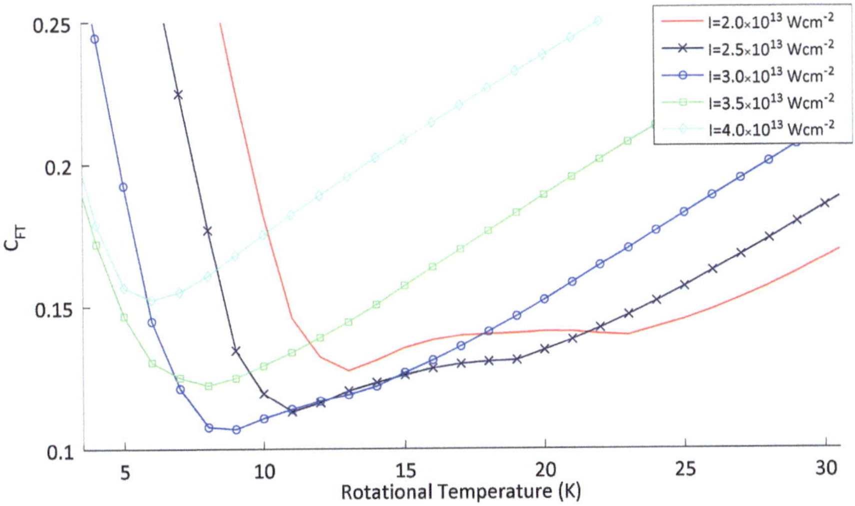

Fig. 5.10 Fourier transform contrast parameter C_{FT} measuring the average difference between experimental and calculated spectral amplitudes of the rotational wavepacket of 10 % N_2 in Ar. The parameter is plotted as a function of rotational temperature for a range of alignment pulse intensities. *Data markers* represent calculated values, whilst the *solid lines* are used to guide the reader's eyes

scaled. For determining the best match with the experimentally acquired spectra, several rotational wavepacket evolutions were simulated for a range of alignment pulse intensities and rotational temperatures of the molecular ensemble. The calculated temporal data was Fourier transformed and the simulated spectra compared to the experimental ones. For a systematic optimisation of the matching, a Fourier transform contrast C_{FT} was defined:

$$C_{FT} = \left\langle \frac{|FT_{exp}| - |FT_{sim}|}{|FT_{exp}| + |FT_{sim}|} \right\rangle \tag{5.16}$$

It calculates the average difference between the spectral amplitudes of the Fourier transforms of the experimentally obtained (FT_{exp}) and simulated (FT_{sim}) data. Minimising this quantity as a function of alignment pulse intensity and rotational temperature thus leads to the best matching simulation. This then enables the retrieval of the temporal evolution of $\langle \cos^2 \theta \rangle$ of the rotational wavepacket and the rotational temperature of the molecular ensemble.

Figure 5.10 displays C_{FT} as a function of rotational temperature for several alignment pulse intensities in the case of $R_{N_2} = 0.1$. Here, the FT contrast is minimised for an alignment pulse intensity of $I = 3 \times 10^{13}$ Wcm^{-2} and a rotational temperature of $T_{rot} = 9$ K. The intensity retrieved from the simulation agrees very well with the estimate obtained in Sect. 4.5.2.

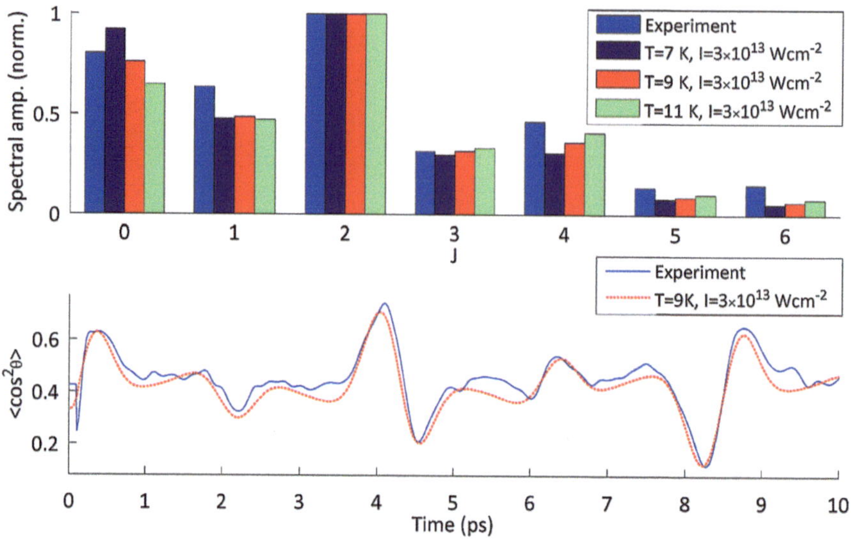

Fig. 5.11 The *upper panel* compares the experimental spectral amplitude distribution for 10% N_2 in Ar to the best matched simulated spectrum with an alignment pulse intensity of $I = 3 \times 10^{13}\,\mathrm{Wcm}^{-2}$ and a rotational temperature of $T = 9\,\mathrm{K}$. Calculated spectra for $T = 7\,\mathrm{K}$ and $T = 11\,\mathrm{K}$ are included as well for illustrating the uncertainty in the rotational temperature extraction. The *lower panel* displays the experimental revival structure and the best matched simulated one

Figure 5.11 then compares the simulated rotational wavepacket to the measured one in the spectral (upper panel) and temporal domain (lower panel). Here, a constant offset and a linear scaling factor are applied to the experimentally obtained revival structure for comparing simulation and experiment in the same plot. A very good agreement between calculation and experiment can be observed in both the spectral and temporal domain. The maximum degree of alignment is achieved at the half revival and was extracted from the simulation yielding $\langle\cos^2\theta\rangle = 0.71$. In order to illustrate the uncertainty of the rotational temperature retrieval, the upper panel displays a comparison of the matching simulation with simulations employing $T_{rot} = 7\,\mathrm{K}$ and $T_{rot} = 11\,\mathrm{K}$ with the same alignment pulse intensity. One can observe that the main features of the experimental spectrum are reproduced by all three simulated spectra. From this comparison, the uncertainty of the extracted rotational temperature is inferred, such that $T_{rot} = 9 \pm 2\,\mathrm{K}$.

The above described procedure was applied to all recorded revival structures. The resulting degrees of alignment characterised as $\langle\cos^2\theta\rangle$ and rotational temperatures T_{rot} are displayed in Table 5.3. The extracted temperatures confirm the qualitative interpretation of the experimental data presented in Sect. 5.5. In the case of N_2, the rotational temperature decreases monotonically with decreasing concentration of the molecular sample within the gas mix. The lowest rotational temperature is thus reached at $R_{N_2} = 0.1$ with 9 K, compared to 24 K for the pure sample.

Table 5.3 Rotational temperatures and maximum molecular alignment quality characterised via $\langle\cos^2\theta\rangle$ obtained from the FT revival analysis

	R_{mol}	1	0.3	0.2	0.1
N_2	T_{rot} (K)	24 ± 4	12 ± 2	10 ± 2	9 ± 2
	$\langle\cos^2\theta\rangle$	0.60	0.67	0.69	0.71
CO_2	T_{rot} (K)	34 ± 3	–	–	9 ± 2
	$\langle\cos^2\theta\rangle$	0.48	–	–	0.64
O_2	T_{rot} (K)	58 ± 2	–	–	37 ± 4
	$\langle\cos^2\theta\rangle$	0.49	–	–	0.58

The results are presented as a function of the concentration R_{mol} of N_2, CO_2 and O_2 in a gas sample mixed with Ar

The enhancement of rotational cooling when seeding the molecule in Ar is largest for CO_2, where the rotational temperature is decreased from 34 to 9 K. This implies a pronounced improvement of the maximum degree of alignment from $\langle\cos^2\theta\rangle = 0.48$ to $\langle\cos^2\theta\rangle = 0.64$.

Given that N_2 and CO_2 are both cooled to 9 K for a 10 % mix, one may conclude that this is the lowest achievable temperature in the supersonic expansion of Ar before the onset of clustering. In the case of O_2, however, the extracted rotational temperatures are significantly higher than the corresponding values for the other molecules. This is the case for the pure and the mixed sample. It is unlikely that the reason for that is inefficient cooling during the supersonic expansion as this would imply a significantly reduced Ar-O_2 collision frequency compared to Ar-N_2 or Ar-CO_2 collisions. A possible reason for the high values of the extracted rotational temperatures may be that the employed theoretical model is not adequate for O_2. In particular, O_2 possesses an open shell structure, with $^3\Sigma_g^-$ ground state. This causes relatively strong spin-spin and spin-orbit interactions that would lead to shifts in the rotational eigenenergies [27]. Such shifts are expected to be small and were not observed in the spectral amplitude distributions obtained from the revival structures. However, the associated effects were not included in the simulations and may be sufficiently significant to render the calculations inappropriate in the case of O_2. This would be a possible explanation for the significantly higher rotational temperatures extracted from the experimental data.

5.7 Conclusion

Laser induced impulsive molecular alignment was applied to the molecules N_2, CO_2 and O_2. The resulting revival structure was probed via Coulomb explosion combined with a TOF detection technique. Even though the molecular alignment distribution could not be retrieved directly from the data, a procedure was developed, which accurately matched numerical simulations to the experimental data for extracting the alignment distribution in form of $\langle\cos^2\theta\rangle$ and the rotational temperature of the molecular ensemble. This was done in the Fourier domain of the temporal evolution of the rotational wavepacket.

The procedure was used for optimising the degree of molecular alignment. For this, the molecule was seeded in Ar as a carrier gas and the revival structure recorded as a function of the mixing ratio. It was found that the rotational temperature decreases with decreasing concentration of the molecule in the mix. The enhanced cooling lead to an improved degree of molecular alignment, which was optimised for a concentration of 10 %. Compared to the pure sample N_2 was cooled from 24 ± 4 to 9 ± 2 K corresponding to an increase of $\langle \cos^2 \theta \rangle$ from 0.60 to 0.71. CO_2 was cooled from 34 ± 3 to 9 ± 2 K improving $\langle \cos^2 \theta \rangle$ from 0.48 to 0.64. For O_2 an enhanced cooling from 58 ± 2 to 37 ± 4 K was observed with an increase of $\langle \cos^2 \theta \rangle$ from 0.49 to 0.58.

The method represents a novel approach for characterising molecular alignment in a TOF spectrometer. So far, frequency spectra of the rotational wavepacket have only been employed for estimating rotational temperatures [19, 28, 29]. The procedure presented in this chapter is the first description of a systematic method for retrieving the rotational temperature and alignment distribution from the FT of the wavepacket evolution. The technique is especially relevant for experiments where the measurement of the revival structure behaves like $\langle \cos^2 \theta \rangle$, but does not allow for a direct extraction of the alignment distribution or rotational temperature. This is the case for TOF spectrometry as in this chapter and high harmonic generation [16], for example.

References

1. H. Stapelfeldt, T. Seideman, Colloquium: aligning molecules with strong laser pulses. Rev. Mod. Phys. **75**, 543 (2003)
2. V. Kumarappan, C.Z. Bisgaard, S.S. Viftrup, L. Holmegaard, H. Stapelfeldt, Role of rotational temperature in adiabatic molecular alignment. J. Chem. Phys. **125**(19), 194309 (2006)
3. F. Filsinger, J. Küpper, G. Meijer, L. Holmegaard, J.H. Nielsen, I. Nevo, J.L. Hansen, H. Stapelfeldt, Quantum-state selection, alignment, and orientation of large molecules using static electric and laser fields. J. Chem. Phys. **131**(6), 064309 (2009)
4. J.J. Larsen, K. Hald, N. Bjerre, H. Stapelfeldt, T. Seideman, Three dimensional alignment of molecules using elliptically polarized laser fields. Phys. Rev. Lett. **85**(12), 2470 (2000)
5. I. Nevo, L. Holmegaard, J.H. Nielsen, J.L. Hansen, H. Stapelfeldt, F. Filsinger, G. Meijer, J. Küpper, Laser-induced 3D alignment and orientation of quantum state-selected molecules, Phys. Chem. Chem. Phys. PCCP **11**, 9912–9918 (2009)
6. R. Torres, R. de Nalda, J.P. Marangos, Dynamics of laser-induced molecular alignment in the impulsive and adiabatic regimes: a direct comparison. Phys. Rev. A **72**(2), 023420 (2005)
7. R.M. Lock, S. Ramakrishna, X. Zhou, H.C. Kapteyn, M.M. Murnane, T. Seideman, Extracting continuum electron dynamics from high harmonic emission from molecules. Phys. Rev. Lett. **108**, 133901 (2012)
8. J. Ortigoso, M. Rodriguez, M. Gupta, B. Friedrich, Time evolution of pendular states created by the interaction of molecular polarizability with a pulsed nonresonant laser field. J. Chem. Phys. **110**, 3870 (1999)
9. T. Seideman, Revival structure of aligned rotational wave packets. Phys. Rev. Lett. **83**, 4971–4974 (1999)

10. O. Ghafur, A. Rouzee, A. Gijsbertsen, W.K. Siu, S. Stolte, M.J.J. Vrakking, Impulsive orientation and alignment of quantum-state-selected NO molecules. Nat. Phys. **5**, 289–293 (2009)
11. M. Leibscher, I.S. Averbukh, H. Rabitz, Molecular alignment by trains of short laser pulses. Phys. Rev. Lett. **90**, 213001 (2003)
12. S. De, I. Znakovskaya, D. Ray, F. Anis, N.G. Johnson, I.A. Bocharova, M. Magrakvelidze, B.D. Esry, C.L. Cocke, I.V. Litvinyuk, M.F. Kling, Field-free orientation of CO molecules by femtosecond two-color laser fields. Phys. Rev. Lett. **103**, 153002 (2009)
13. J.G. Underwood, B.J. Sussman, A. Stolow, Field-free three dimensional molecular axis alignment. Phys. Rev. Lett. **94**, 143002 (2005)
14. K.F. Lee, D.M. Villeneuve, P.B. Corkum, A. Stolow, J.G. Underwood, Field-free three-dimensional alignment of polyatomic molecules. Phys. Rev. Lett. **97**, 173001 (2006)
15. F. Rosca-Pruna, M. Vrakking, Experimental observation of revival structures in picosecond laser-induced alignment of i_2. Phys. Rev. Lett. **87**, 1–4 (2001)
16. T. Kanai, S. Minemoto, H. Sakai, Quantum interference during high-order harmonic generation from aligned molecules. Nature **435**(7041), 470–474 (2005)
17. R. Torres, T. Siegel, L. Brugnera, I. Procino, J.G. Underwood, C. Altucci, R. Velotta, E. Springate, C. Froud, I.C.E. Turcu, M.Y. Ivanov, O. Smirnova, J.P. Marangos, Extension of high harmonic spectroscopy in molecules by a 1300 nm laser field. Opt. Express **18**(3), 3174–3180 (2010)
18. V. Renard, M. Renard, A. Rouzée, S. Guérin, H. Jauslin, B. Lavorel, O. Faucher, Nonintrusive monitoring and quantitative analysis of strong laser-field-induced impulsive alignment. Phys. Rev. A **70**, 1–9 (2004)
19. D. Pinkham, R.R. Jones, Intense laser ionization of transiently aligned co. Phys. Rev. A **72**(2), 023418 (2005)
20. G. Herzberg, *Molecular Spectra and Molecular Structure: Spectra of Diatomic Molecules*, 2nd edn. (Krieger Pub Co, Malabar, 1989)
21. P. Atkins, R. Friedman, *Molecular Quantum Mechanics*, 4th edn. (Oxford University Press, USA, 2005)
22. M. Spanner, Field-Free Alignment and Strong Field Control of Molecular Rotors. Ph.D. thesis, University of Waterloo, 2004
23. H. Haken, *Molecular Physics and Elements of Quantum Chemistry: Introduction to Experiments and Theory* (Springer, Berlin, 2010)
24. M. Hillenkamp, S. Keinan, U. Even, Condensation limited cooling in supersonic expansions. J. Chem. Phys. **118**, 8699–8705 (2003)
25. R. Campargue, Progress in overexpanded supersonic jets and skimmed molecular beams in free-jet zones of silence. J. Phys. Chem. **88**, 4466–4474 (1984)
26. H.R. Murphy, D.R. Miller, Effects of nozzle geometry on kinetics in free-jet expansions. J. Phys. Chem. **88**(20), 4474–4478 (1984)
27. J.M. Brown, A. Carrington, *Rotational Spectroscopy of Diatomic Molecules* (Cambridge University Press, Cambridge, 2003)
28. K. Yoshii, G. Miyaji, K. Miyazaki, Dynamic properties of angle-dependent high-order harmonic generation from coherently rotating molecules. Phys. Rev. Lett. **101**, 183902 (2008)
29. K. Yoshii, G. Miyaji, K. Miyazaki, Measurement of molecular rotational temperature in a supersonic gas jet with high-order harmonic generation. Opt. Lett. **34**, 1651–1653 (2009)

Chapter 6
Multichannel Contributions in Nonsequential Double Ionisation of CO_2

In this chapter the molecular alignment dependence of recollision induced double ionisation is studied. In particular, the orbital symmetry of the contributing molecular states is encoded into the angular dependent signature of the process. For resolving this signature, impulsive molecular alignment was used to study the double ionisation event in the molecular frame. This enabled the identification of the specific ionic states in the underlying mechanism. Using CO_2 as an example, the wavelength of the strong driving laser field was increased from 800 to 1,350 nm at 2×10^{14} Wcm^{-2}. This established nonsequential double ionisation (NSDI) as the exclusive mechanism for CO_2^{++} formation. Under these conditions the recollision event was fully characterised by measuring the shape of the recolliding electron wavepacket and the inelastic electron-ion recollision cross section. The obtained results unambiguously reveal the contribution from both the ionic ground and first excited state to the NSDI mechanism in CO_2. Additionally, the experiment represents the first angularly resolved electron-ion collision measurement in CO_2, albeit over a finite collision energy range.

6.1 Background

The control of strong field induced dynamics in small molecules on the femtosecond time scale continues to be a fruitful branch of research; also for establishing the experimental and theoretical tools required to go towards ever shorter time scales and more complex molecular systems. The strong field regime can be characterised by the onset of field ionisation. While the ionisation channel itself can partially be selected via the laser intensity and polarisation [1, 2], it also provides an ideal playground for the strong field physicist by triggering electronic and nuclear dynamics in the remaining parent ion. These have previously been probed through laser driven electron recollision [3] in the form of high harmonic generation (HHG). This has lead to milestone experiments tracking down nuclear motion in molecules [4], structural changes during chemical reactions [5], the tomographical imaging of molecular orbitals [6] and to identify specific ionisation channels and track subsequent hole dynamics [2, 7], to name but a few examples. Here, a key advantage of

M. Oppermann, *Resolving Strong Field Dynamics in Cation States of CO_2 via Optimised Molecular Alignment*, Springer Theses, DOI: 10.1007/978-3-319-05338-7_6,

HHG is that it is exclusively driven by recollision. This is not the case for inelastic recollision phenomena, such as nonsequential double ionisation (NSDI), which competes with the direct laser field induced process of sequential double ionisation (SDI). This competition with purely laser induced channels typically complicates the control of inelastic recollision effects and so inhibits the extraction of information on molecular structure and dynamics.

In the experiments presented in this chapter, this problem was overcome by employing a long wavelength driving laser pulse to establish inelastic recollisions as the exclusive mechanism for double ionisation (DI). Under these conditions the ellipticity and polarisation dependence of NSDI in CO_2 was studied in the molecular frame. The results enable the identification of specific intermediate molecular states in the underlying NSDI mechanism. This can be regarded as the first experiment resolving multiple channels in molecular NSDI and providing significant progress for improving inelastic recollisions as a probe for molecular dynamics.

6.1.1 Strong Field Ionisation

It is usually assumed that tunneling is the dominating ionisation mechanism, once the tunneling frequency significantly exceeds the driving laser frequency. This is reflected in the Keldysh adiabaticity parameter fulfilling $\gamma = \frac{\omega_{laser}}{\omega_{tunnel}} < 1$ [8]. The tunnel-ionisation step also depends on the molecular structure, as described in the molecular Ammosov-Delone-Krainov (MO-ADK) theory [1]. In this framework, the essential features of the angular ionisation rate are obtained under the assumption that the ionisation probability scales with the electronic density profile of the respective molecular orbital along the ionising laser polarisation direction. Due to the exponential decay of the MO-ADK rate with the ionisation potential I_p, ionisation from the highest occupied molecular orbital (HOMO) typically dominates. However, it has been observed that the influence of lower lying orbitals cannot be excluded a priori [2, 9–11]—especially not in CO_2. In a Koopmans type picture this implies that excited states in the parent ion can be reached through ionisation from lower lying orbitals. In aligned CO_2, the probability of reaching an excited state via this channel then scales like the MO-ADK rate for the associated orbital. The relative strength of this channel can thus be controlled by the laser polarisation with respect to the molecular axis. It should be added however, that experimental ionisation studies in aligned CO_2 show quantitative discrepancies with the MO-ADK model [12], suggesting significant shortcomings (alternative approaches and modifications can be found in [13–17] for example).

6.1.2 Nonsequential Double Ionisation

Double ionisation may proceed through tunnel-ionisation of the parent ion, where the associated ADK rate simply employs the second ionisation potential I_p^+. In such a sequential process, the two ionisation events are independent of each other. This

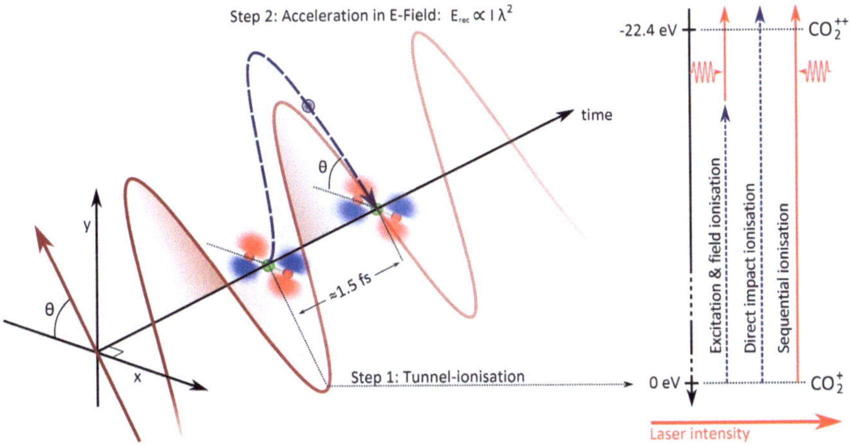

Fig. 6.1 Strong field double ionisation channels in CO_2. NSDI is driven by electron recollision, where the second ionisation potential can either be overcome by recollision excitation followed by field ionisation or by direct impact ionisation. Sequential double ionisation is purely field induced and independent of the first ionisation step

channel dominates when the intensity passes the saturation threshold, such that the potential barrier of the molecule is suppressed and the ground state is depleted. It is well known however that sufficiently below the saturation regime, double ionisation yields are enhanced and in some cases even dominated by electron recollision [18]. This channel has been termed non-sequential double ionisation (NSDI), highlighting the correlation of the two ionisation events and ejected electron wavepackets. The maximum recollision energy the free electron can acquire in the laser field is $E^r \approx 3.17U_p$ where $U_p \propto I\lambda^2$ is its ponderomotive energy in a laser field with intensity I and wavelength λ [3]. An inelastic recollision may then lead to electron impact excitation or ionisation. Generally, low intensities at a fixed wavelength imply a shift towards the multiphoton regime, favouring excitation over direct impact ionisation. Starting from a low intensity, recollision excitation with subsequent field ionisation (RESI) [19] will dominate until the recollision energy exceeds the second ionisation potential I_p^+. Here, direct impact ionisation becomes possible [20, 21] (see Fig. 6.1). For further increase of the intensity, sequential ionisation becomes more likely until the saturation regime is reached, where it dominates. The ionisation potentials relevant for this experiment are $I_p = 15.8\,\text{eV}$ and $I_p^+ = 27.6\,\text{eV}$ for Ar [22] and $I_p = 13.8\,\text{eV}$ and $I_p^+ = 22.4\,\text{eV}$ for CO_2 [23].

Intensity Dependence

NSDI in CO_2 has previously been observed in an intensity regime of $0.5-2 \times 10^{14}\,\text{Wcm}^{-2}$ at 800 nm via the intensity dependence of the ratio X^{2+}/X^+, where X refers to the molecule [24]. For a nonsequential process X^+ is an intermediate state, such that the intensity dependence of the above ratio should be weak. For sequential ionisation on the other hand, X^+ is a precursor state and hence the corresponding ratio should have a strong intensity dependence as this would lead to a depletion of the X^+ yield along with an increase of the X^{2+} yield.

Ellipticity Dependence

While high precision intensity dependent studies are very cumbersome due to focal volume averageing, recollision induced channels can also be unambiguously identified via their ellipticity dependence. In particular, pure recollision processes like HHG are fully suppressed for 800 nm fields with ellipticities $\epsilon > 0.3$ [25], where the ellipticity is defined as the electric field component ratio along the y- and x-axis (see also Fig. 6.5).

The ellipticity dependence of recollision induced phenomena in molecules also encodes the structure of the ionised molecular orbital. When the tunneling ionisation step takes place along a nodal plane in the molecular orbital, the ejected free electron wavepacket has a node on axis in coordinate space. This is due to destructive quantum interference between parts of the tunneled electron wavepackets from orbital parts with opposite parity. The node in coordinate space is associated with a node on axis in momentum space, which implies that most of the classical trajectories associated with the free electron wavepacket possess nonzero transverse momentum. In a linearly polarized driving laser field, this means that most electrons are drifting away, perpendicular to the laser polarisation direction during their propagation in the continuum and thus miss the parent ion upon return. However, in an elliptically polarized field, the initial transverse momentum can be compensated for by the transverse component of the laser electric field. This leads to a maximum number of recollision events for a nonzero ellipticity, such that a local minimum or dip is observed for $\epsilon = 0$. Here, the ellipticity curve maps out the transverse velocity distribution of the recolliding electron wavepacket. This effect has been observed in NSDI of C_6H_6 [26].

Polarisation Dependence

Studies on the polarisation dependence of NSDI in the molecular frame are still scarce (see Sect. 2.4.2). In the case of CO_2, differential ion recoil momentum (COLTRIMS) studies have revealed the angular dependence of recollision induced dissociative double ionization [27] (see Fig. 2.17 for details). The latter approach is used mostly to study repulsive final states from which the molecular alignment and thus molecular frame can be reconstructed. The study of metastable doubly charged ions requires additional active molecular alignment in the laboratory frame. This has thus far only been realised in the case of N_2, but only parallel and perpendicular rates with respect to the molecular axis were compared [28]. However, additional theoretical efforts [11, 29, 30] have indicated that the recollision excitation cross section of the parent ion roughly scales like its electron density profile. This suggests that the polarisation dependence of NSDI in the molecular frame may reveal the involvement of excited ionic states in this process.

Wavelength Dependence

The wavelength dependence of NSDI is qualtiatively similar to its intensity dependence, but quantitatively stronger due to the quadratic scaling of U_p with the wavelength. Increasing the wavelength thus has several advantages when studying inelastic recollision processes. Firstly, increasing the wavelength implies higher recollision energies. This opens channels that require large excitation energies, such as overcoming the second ionisation potential I_p^+ for NSDI. Secondly, the Keldysh parameter decreases. This favours tunneling over multiphoton ionisation, which implies that recollision induced phenomena are favoured over direct laser induced processes. In the case of NSDI this means that a long driving laser field at low intensity may lead to recollision dominating the double ionisation yield.

6.2 Analytical Model for Estimating the Visibility of Quantum Interference

In the previous section it was pointed out that quantum interference in recollision induced phenomena can be used as a probe for nodal planes and thus molecular structure of the ionised orbital. Here, the visibility of the associated dip in the ellipticity dependence of NSDI in an experiment that detects doubly charged ions relies on three conditions: (1) NSDI must be the exclusive double ionisation process, (2) the electron wavepacket must have a node centred at zero perpendicular momentum, and (3) the formation of the detected doubly charged ions must be dominated by recollision events involving electron wavepackets possessing a node on axis in momentum space.

(1) is required because sequential double ionisation does not proceed through recollision and thus cannot display quantum interference. Any contributions from sequential channels will thus wash out a possible dip in the ellipticity dependence of the doubly charged ion. This point was addressed in the experiment by employing a long driving laser wavelength of 1,350 nm. (2) is fulfilled if the tunnel ionisation (TI) step takes place along the node of the ionized orbital. For CO_2 this is the case for ionisation at $\theta_L = 0°$ and $\theta_L = 90°$, where θ_L is the angle between the molecular axis and the linear laser polarisation. (3) is, however, the crucial requirement for resolving quantum interference in NSDI in an experiment. This is because the detected NSDI yield results from a molecular ensemble with a continuous alignment distribution with respect to the ionising laser polarization. Yet only those molecules with a nodal plane aligned along the laser polarisation will display quantum interference. For the visibility of this effect in an experiment, this has two implications. Firstly, the visibility of the ellipticity dip increases with increasing width of the involved nodal planes. Secondly, the visibility improves with increasing angular confinement of the probed nodal plane within a solid angle around the laser polarisation. In the case of the HOMO of CO_2, the former point implies that quantum interference is significantly more likely to be visible at $\theta_L = 90°$ compared to $\theta_L = 0°$.

The latter point then implies that a significantly higher degree of molecular alignment is required to observe the nodal plane at $\theta_L = 0°$.

In order to estimate the visibility of quantum interference in an ellipticity measurement under the employed experimental conditions, Sébastien Weber developed an analytical model for the ellipticity dependence of electron recollision with the HOMO of CO_2. The model is based on an ansatz proposed by Misha Ivanov. The obtained results were then convolved with the alignment distribution measured in the experiments presented in this chapter (see Sect. 6.3). His calculations are briefly presented in this section, which is based on [31].

6.2.1 Single Molecule

As all relevant steps in the recollision event are confined to the laser polarisation plane, the model is reduced to two dimensions. The geometry of the HOMO and the laser field is depicted in Fig. 6.1, where the molecule is assumed to be perfectly aligned. Here, its molecular axis is aligned along the coordinate x-axis and laser polarisation is aligned at an angle θ_L (θ in the figure) with respect to the molecular axis. For an arbitrary value of θ_L, the outgoing wavepacket is a superposition of two wave functions that describe the transverse momentum distribution of the ejected electron: a Gaussian without a node (ψ^+) and a Gaussian with a node at zero transverse momentum (ψ^-). ψ^+ results from ionisation along an axis through the molecular orbital without a nodal plane and ψ^- is obtained when the molecule is ionised along a nodal plane. Mathematically, $\psi^\pm$ are obtained via the spatial Fourier transform of the ionised Dyson orbital in the direction perpendicular to the ionising laser polarisation axis (for a detailed derivation see [31]). The ellipticity dependence of NSDI within this model is then obtained from the transverse momentum distribution of the tunneled electron wavepacket and the requirement for recollision in an elliptical field [31]:

$$v_\perp \approx \epsilon \frac{E_L}{\omega_L} \tag{6.1}$$

where $v_\perp$ is the transverse momentum and E_L and ω_L the laser electric field amplitude and frequency, respectively. This leads to ellipticity curves $P^\pm(\epsilon)$ that describe the NSDI rate as a function of laser pulse ellipticity. In particular, ψ^+ results in a Gaussian ellipticity dependence $P^+(\epsilon)$ and ψ^- leads to a dip at zero ellipticity and a maximum NSDI rate for nonzero ϵ described by $P^-(\epsilon)$:

$$P^+(\epsilon) = \frac{1}{N^+} e^{-\left(\epsilon/\epsilon_0^+\right)^2} \tag{6.2}$$

$$P^-(\epsilon) = \frac{\epsilon^2}{N^-} e^{-\left(\epsilon/\epsilon_0^-\right)^2} \tag{6.3}$$

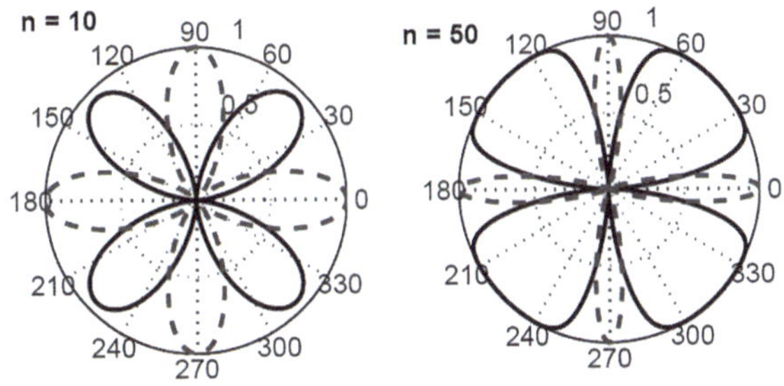

Fig. 6.2 Relative probabilities w^- for the tunneled electron wavepacket to possess a node (*dashed line*) and w^+ without a node (*solid line*) as a function of laser polarisation angle with respect to the molecular bond. The *left panel* displays a behaviour associated with broad nodal planes at $\theta_L = 0°$ and $\theta_L = 90°$, whilst the right panel displays very narrow nodal planes at the same angular positions (Reprinted with permission from [32]. Copyright (2013) by the American Physical Society)

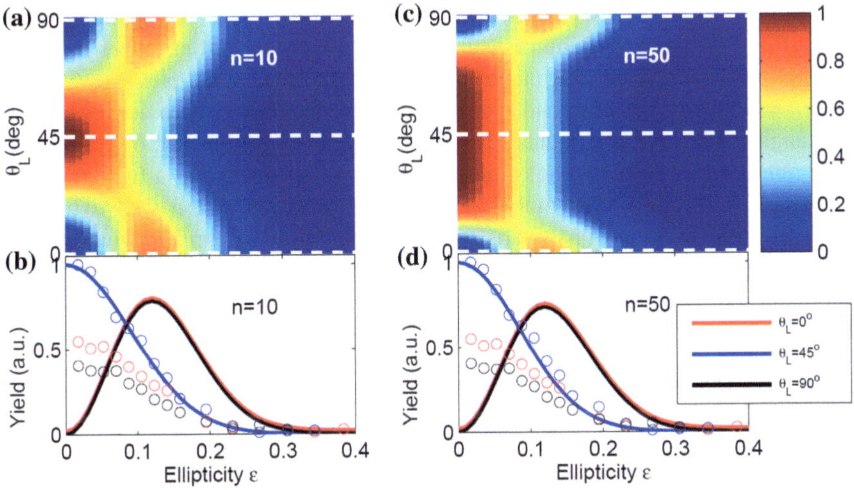

Fig. 6.3 NSDI yield from a single, perfectly aligned molecule as a function of laser polarization angle θ_L with respect to the molecular axis and laser ellipticity ϵ. **a** and **c** display the full 2-dimensional maps for the two limiting cases of the nodal structure. **b** displays the lineouts from **a** and **d** shows the lineouts obtained from **c**. For comparison, the experimental data from Fig. 6.11 is displayed in **b** and **c** as well (*open circles*). This data will be discussed in Sect. 6.6 (Reprinted with permission from [32]. Copyright (2013) by the American Physical Society)

where $N^{\pm}$ are normalisation constants and $\epsilon_0^{\pm}$ constants that determine the width of the ellipticity curves. Here, ϵ_0^+ was obtained from the experimental measurement for $\Delta\epsilon_{1350}$. The relationship $\epsilon_0^- = 0.7\epsilon_0^+$ was then obtained numerically via the condition that $P^-(\epsilon) = P^+(\epsilon)$ for $\epsilon > 0.3$, where both distributions are expected to vanish.

The relative weights of ψ^+ and ψ^- in their superpositon now depend on the ionisation direction θ_L with respect to the molecular orbital. In particular, ψ^+ dominates when TI proceeds along a maximum of the electron density of the ionised orbital, whilst ψ^- dominates when this takes place along one of its nodal planes. Furthermore, the relative contribution of ψ^- as a function of θ_L crucially depends on the width of the nodal plane. This was taken into account by modelling the orbital structure via parametric functions $w^\pm$ that serve as relative angular probabilities for the superposition of $\psi^\pm$:

$$w^+(\theta_L) = 1 - w^-(\theta_L) \tag{6.4}$$

$$w^-(\theta_L) = |\cos\theta_L|^n + |\sin\theta_L|^n \tag{6.5}$$

where n determines the width of the nodes at $\theta_L = 0°$ and $\theta_L = 90°$. This then results in a total ellipticity dependence of:

$$P_{\text{tot}}(\theta_L, \epsilon) = w^-(\theta_L)P^-(\epsilon) + w^+(\theta_L)P^+(\epsilon) \tag{6.6}$$

Because the HOMO of CO_2 has a narrow node at $\theta_L = 0°$ and a much wider one at $\theta_L = 90°$, two limiting cases for the relative angular probabilities $w^\pm$ were investigated. The first limiting case (n = 10) displays broad nodal planes at $\theta_L = 0°$ and $\theta_L = 90°$, whilst the second limiting case (n = 50) shows sharp nodal planes at both angles. The resulting angular distributions are displayed in Fig. 6.2.

Figure 6.3 then displays the resulting ellipticity dependence as a function of θ_L for a single, perfectly aligned molecule with the nodal structure from the two limiting cases. The full 2-dimensional maps show that the ellipticity dip is angularly more confined in the limiting case with the sharper nodal planes (n = 50). However, when examining the lineouts at $\theta_L = 0°$ and $\theta_L = 90°$, the two limiting cases lead to the same result. This is because the width of the nodal plane does not matter for a perfectly aligned molecule. Note that in these cases, the $P^-(\epsilon, \theta_L)$ contribution fully dominates, so the ellipticity curve displays zero NSDI yield at $\epsilon = 0$.

6.2.2 Convolution With Molecular Alignment Distribution

For estimating the visibility of quantum interference in the experiment presented in this chapter, the results displayed in Fig. 6.3 had to be convolved with the experimentally measured molecular alignment distribution (for details on the procedure see [31]). This effectively washes out the node contribution $P^-(\epsilon, \theta_L)$ as it is averaged over a solid angle related to the alignment distribution. The results of this convolution procedure are displayed in Fig. 6.4. Assuming that only the HOMO orbital of CO_2 is contributing to the NSDI mechanism, the results now allow for two key conclusions concerning the visibility of quantum interference in our experiment. Firstly, the nodal plane at $\theta_L = 90°$ is expected to be encoded in the ellipticity dependence of the NSDI yield. Due to the orbital structure of the HOMO, it is appropriate to assume a large width of the nodal plane (n = 10) perpendicular to the molecular axis. In this case a dip at $\epsilon = 0$ is expected to be visible. However, even if one assumes a sharper nodal plane (n = 50), the full width at half maximum (FWHM) of the resulting ellipticity

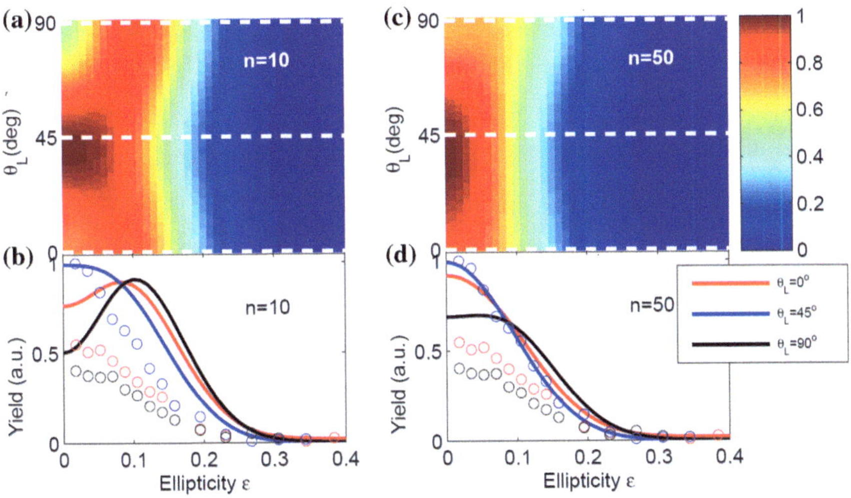

Fig. 6.4 NSDI yield from an ensemble of aligned molecules as a function of laser polarization angle θ_L with respect to the molecular axis and laser ellipticity ϵ. For this, the results displayed in Fig. 6.3 are convolved with the experimentally measured alignment distribution and angular ionization rate. **a** and **c** display the full 2-dimensional maps for the two limiting cases of the nodal structure. **b** displays the lineouts from **a** and **d** shows the lineouts obtained from **c**. For comparison, the experimental data from Fig. 6.11 is displayed in **b** and **c** as well (*open circles*). This data will be discussed in Sect. 6.6 (Reprinted with permission from [32]. Copyright (2013) by the American Physical Society)

curve would significantly differ from the one at $\theta_L = 45°$, even though the analytical model is not very accurate in this angular range, because the HOMO of CO_2 is not symmetric with respect to this angle. Secondly, the visibility of the nodal plane at $\theta_L = 0°$ is significantly reduced compared to $\theta_L = 90°$, to the extent that it is very unlikely to be observed in the experiment presented in this chapter. This is due to the much sharper nodal plane parallel to the molecular bond. For this, the limiting case (n = 50) is the more appropriate one. However, the degree of molecular alignment is not sufficient to resolve the nodal plane in the associated ellipticity curve. In fact, it only marginally differs from the ellipticity dependence at $\theta_L = 45°$, where the laser polarisation is directed along a maximum of the electronic density profile.

6.3 Experiment

The aim of this experiment was to study the molecular structure dependence of NSDI in CO_2 and to infer the participating molecular states via their orbital symmetries. For achieving this, all laser parameter dependences introduced previously were studied and the experiment was divided into three stages. Firstly, NSDI in Ar was investigated for benchmarking the experiment. Secondly, NSDI was established as the exclusive

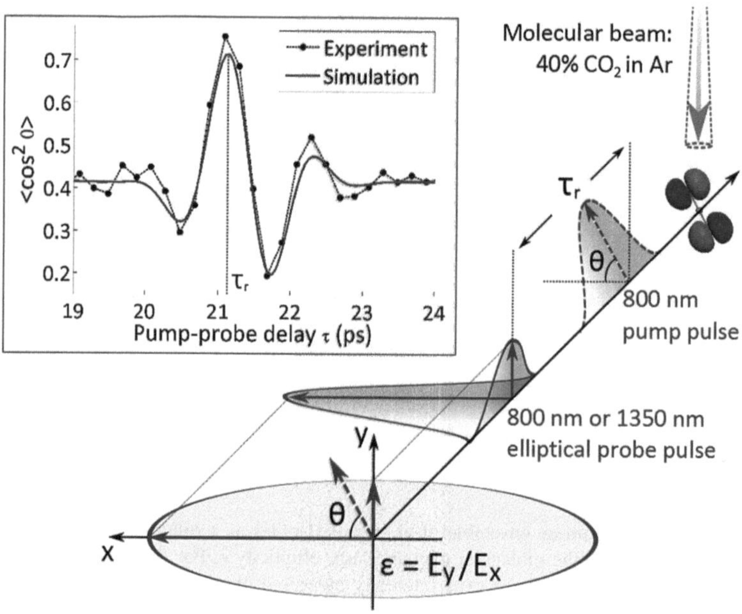

Fig. 6.5 Pump-probe scheme for investigating the ellipticity and polarization dependence of NSDI in impulsively aligned CO_2. Ions were detected via a TOF spectrometer operated in Wiley-McLaren mode. The *inset* displays a half-revival of the rotational wavepacket induced by the pump pulse under the employed experimental conditions (Reprinted with permission from [32]. Copyright (2013) by the American Physical Society)

mechanism for DI by conducting intensity and ellipticity scans at 800 and 1,350 nm. Thirdly, the recollision event was characterised in the molecular frame via impulsive alignment. Here, the shape of the recolliding electron wavepacket was recorded via the ellipticity dependence of CO_2^{++} and the recollision cross section extracted via its polarisation dependence.

6.3.1 Setup and Procedure

The details of the pump-probe beamline employing either 800 or 1,350 nm probe pulses were presented in Sect. 4.3.2. A schematic drawing of the experimental scheme is displayed in Fig. 6.5. Here, the quarter-wave plate was placed in the probe and the half-wave plate in the alignment arm. For ellipticity scans the quarter-wave plate changed the probe ellipticity ϵ. The half-wave plate was adjusted to keep the alignment pulse polarisation and thus molecular alignment distribution parallel to the major axis of the probe polarisation ellipse. For polarisation scans the alignment polarisation and thus molecular alignment distribution is rotated with respect to the fixed linear probe polarisation. Here, θ denotes the angle between probe and

Table 6.1 Probe laser parameters and resulting peak (I_0) and average intensities (I_{av}) employed for the experimental measurements

λ_p (nm)	ΔE_p (μJ)	$\Delta\tau_p$ (fs)	Δw_p (μm)	I_0 (Wcm^{-2})	I_{av} (Wcm^{-2})
800	120	30	41	3.0×10^{14}	$(2 \pm 1) \times 10^{14}$
	190	30	41	4.8×10^{14}	$(3 \pm 2) \times 10^{14}$
1350	330	30	61	3.8×10^{14}	$(2 \pm 1) \times 10^{14}$

Table 6.2 Alignment laser parameters for each probe wavelength λ_p, resulting peak (I_0) and average intensities (I_{av}) and degree of molecular alignment at the half revival of the induced rotational wavepacket

λ_p (nm)	ΔE_a (μJ)	$\Delta\tau_a$ (fs)	Δw_a (μm)	I_0 (Wcm^{-2})	I_{av} (Wcm^{-2})	$\langle\cos^2\theta\rangle$
800	150	70	57	8.4×10^{13}	$(5 \pm 4) \times 10^{13}$	0.54
1,350	170	75	53	1.0×10^{14}	$(5 \pm 5) \times 10^{13}$	0.71

alignment polarisation. The required calibration of the wave plates was described in detail in Sect. 4.5.4. The probe laser parameters used in the conducted measurements are displayed in Table 6.1. The resulting peak (I_0) and average intensities (I_{av}) were determined according to the procedure explained in Sect. 4.5.2. Table 6.2 then shows the alignment pulse parameters and obtained alignment distributions at the probed half revival (for details on the implementation and characterisation of impulsive molecular alignment see Chap. 5). Note that despite the relatively high peak intensities of the alignment pulses, no double ionisation signal was present from the alignment pulse alone. Furthermore the single ionisation yield from the alignment pulse was more than one order of magnitude lower than the corresponding signal from any of the employed probe pulses.

The molecular beam was operated at a repetition rate of 47 Hz and an effective valve opening time of 30 μs. Throughout all measurements, the backing pressure was kept at 2.3 ± 0.1 bar and the opening distance of the valve was set to maintain a background pressure in the source chamber of 3×10^{-5} mbar. This resulted in a background pressure in the interaction chamber below 8×10^{-8} mbar during operation of the gas jet. The gas sample was a binary mix consisting of 40 % CO_2 in Ar. The preparation of a mixed gas sample was presented in Sect. 4.3.4. The reason why the previously found optimised mixing ratio of 10 % (see Chap. 5) was not used is that the background pressure in the source chamber had been improved after the molecular alignment experiments presented in Chap. 5. This caused an increased rotational cooling in the supersonic gas expansion and thus clustering in the 10 % mix. Increasing the mixing ratio was thus required for avoiding this effect.

The TOF spectrometer was operated under Wiley McLaren space focussing conditions as presented in Sect. 4.2. All measurements were performed in single shot acquisition mode for obtaining TOF spectra with high signal to noise and thus dynamic range (for details see Sect. 4.2.2). This was necessary as signals with count rates differing by several orders of magnitude had to be monitored simultaneously within one TOF spectrum. The number of single shot spectra used for one data point was

Table 6.3 Number of recorded single shot spectra comprising one average spectrum for a data point depending on the probe wavelength, intensity and varied scan parameter

λ_p (nm)	Scan parameter	I_{av} (Wcm^{-2})	Shots per data point
800	I_0	–	8,000
	ϵ, θ	$(2 \pm 1) \times 10^{14}$	8,000
	ϵ, θ	$(3 \pm 2) \times 10^{14}$	5,000
1,350	I_0	–	10,000
	ϵ	$(2 \pm 1) \times 10^{14}$	20,000
	θ	$(2 \pm 1) \times 10^{14}$	15,000

Here, I_0 refers to the peak intensity, ϵ to the ellipticity and θ to the polarisation angle

chosen according to the number of recorded double ionisation (DI) events per laser shot. This is because at high laser intensities, more DI events take place and the acquired number of shots for obtaining a given signal to noise ratio can be reduced. Additionally, significantly lower count rates were found for the 1,350 nm compared to the 800 nm probe. The employed number of shots per data point for the different measurements are displayed in Table 6.3.

Intensity scans were performed by varying the probe pulse energy and thus peak intensity I_0 via the variable attenuator. For both employed probe wavelengths, the scans were performed with the alignment pulse blocked and thus probing an unaligned sample. Ellpticity scans in aligned molecules were performed by rotating the quarter-wave plate in the probe arm. Here, scans were performed for three orientations of the molecular ensemble with respect to the major axis of the probe polarisation ellipse: $\theta = 0°, 45°, 90°$ Polarisation scans in aligned molecules were performed by rotating the half-wave plate in the alignment arm with respect to the linear probe polarisation.

6.3.2 Data Analysis

Figure 6.6 displays a TOF spectrum obtained from a linearly polarised, 1,350 nm probe pulse at 2×10^{14} Wcm^{-2}. The single and double ionisation yields of Ar and CO_2 were obtained by numerically integrating the associated peaks. As an example, the integration bounds are indicated as dashed red lines for Ar^{++} and CO_2^{++} in the figure inset. The DC offset of each peak within the spectrum was obtained by calculating the mean of a section of the signal height next to the peak. These sections are indicated via red dotted lines on the left of the Ar^{++} and CO_2^{++} peaks in the figure inset.

The Ar^+ signal was used to account for jitter in TOF and laser intensity fluctuations between data points. The jitter was corrected by extracting the peak positions of the Ar^+ signal. With the first data point in a scan fixing the peak position, subsequent spectra were shifted in TOF such that their associated Ar^+ peak positions coincided with the reference from the first data point. For ellipticity scans, the Ar^+ yield $A(\epsilon, \theta)$ is independent of θ due to the spherical symmetry of the atom. Therefore, intensity

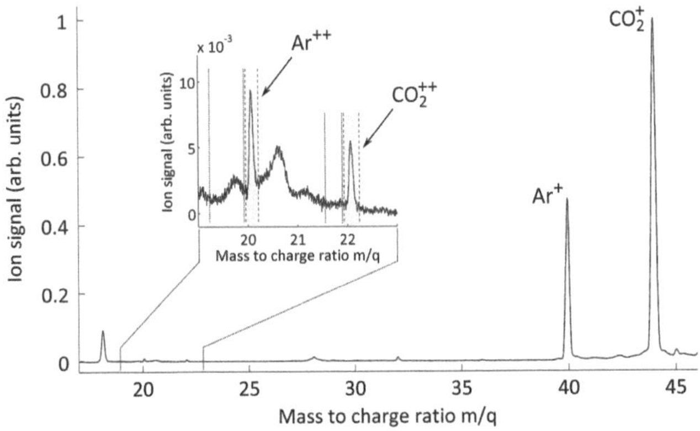

Fig. 6.6 TOF mas spectrum obtained from a mixed gas sample consisting of 40 % CO_2 in Ar. The probe was a linearly polarised, 1,350 nm, 30 fs pulse at 2×10^{14} Wcm^{-2}

drifts between scans for different values of θ were accounted for by multiplying the yields of all ion species with a correction factor given by $C(\epsilon, \theta) = \langle A(\epsilon)\rangle / A(\epsilon, \theta)$, where $\langle A(\epsilon)\rangle = (A(\epsilon, 0°)+A(\epsilon, 45°)+A(\epsilon, 90°))/3$. For polarisation scans, the Ar^+ yield $A(0, \theta)$ is expected to be constant again so the same correction factor $C(0, \theta)$ was used to account for intensity drifts between data points within a polarisation scan. Note that now $\langle A(0)\rangle$ is the mean over all values θ probed in the polarisation scan.

6.4 NSDI in Ar

It is well known that NSDI becomes the dominant DI mechanism at intensities around 2×10^{14} Wcm^{-2} in 800 nm driving laser fields. This makes it a suitable target for benchmarking the experimental setup by studying this behaviour as a function of the available laser parameters. In particular, the onset of recollision induced DI will be characterised via the intensity dependence of Ar^{++} and its ellipticity dependence. Here, the two average probe intensities for 800 nm are $I_{800}^{(lo)} = 2 \times 10^{14}$ Wcm^{-2} and $I_{800}^{(hi)} = 3 \times 10^{14}$ Wcm^{-2}. In the case of 1,350 nm $I_{1350}^{(lo)} = 2 \times 10^{14}$ Wcm^{-2} was used.

6.4.1 Results

Figure 6.7 displays the yields of Ar^+, Ar^{++} and their ratio Ar^{++}/Ar^+ as a function of peak intensity I_0 for an 800 nm (left panel) and a 1,350 nm probe pulse (right panel). The dotted vertical lines denote the peak intensities employed for conducting ellipticity and polarisation measurements. The corresponding average intensities are

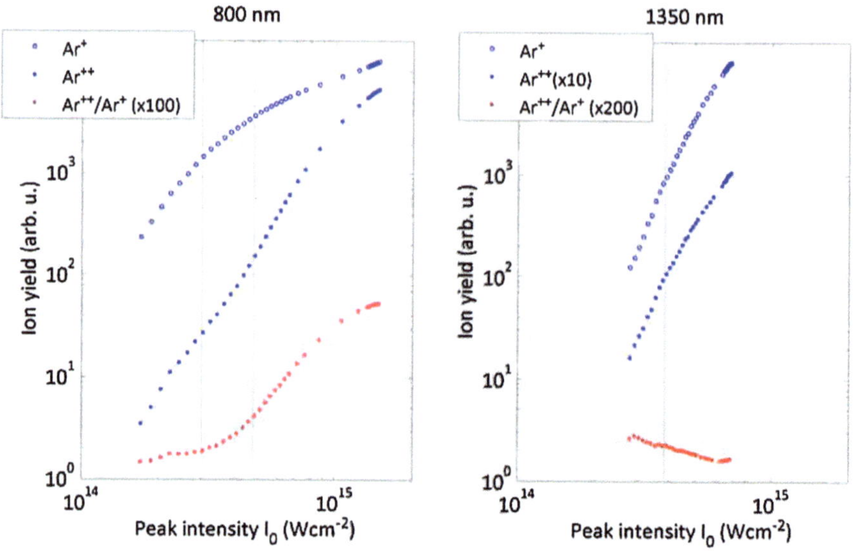

Fig. 6.7 Ar^+, Ar^{++} and their ratio Ar^{++}/Ar^+ as a function of peak intensity I_0 for 30 fs probe pulses with a wavelength of 800 nm (*left panel*) and 1,350 nm (*right panel*). The *dotted vertical lines* denote the peak intensities employed for conducting ellipticity and polarisation measurements

denoted as $I_{800}^{(hi)}$, $I_{800}^{(lo)}$ and $I_{1350}^{(lo)}$, as discussed in Sect. 6.3.1. In the case of 800 nm radiation, the Ar^{++} displays an inflection point at a peak intensity of about $4 \times 10^{14}\,\mathrm{Wcm}^{-2}$. Following the DI rate from high to low intensities, this means that its curvature changes at the inflection point. This cannot be described by employing an exponential tunneling-ionisation rate as in the ADK model (see Eq. 2.6). Therefore, the DI rate must be enhanced by an ionisation process other than tunneling for intensities lower than this inflection point. The shape of the resulting curve is usually referred to as a 'knee' structure. Below the inflection point, the slope of the DI rate is similar to the single ionisation (SI) rate. This becomes clear when examining the ion yield ratio Ar^{++}/Ar^+. Starting at the highest probed peak intensity, the slope of the curve is almost flat. With decreasing intensity, one then observes a decreasing ion yield ratio. This indicates that the Ar^{++} yield decreases faster than the Ar^+ yield with decreasing intensity. When reaching the inflection point, the slope begins to decrease and becomes flat below the peak intensity for $I_{800}^{(lo)}$. In the 1,350 nm case, the Ar^{++} does not display an inflection point in the probed intensity regime. However, over the entire regime the slope of the DI rate is very similar to the SI rate. This manifests itself in a nearly flat ion yield ratio, which increases slightly with decreasing intensity. This implies that the Ar^{++} yield decreases slower than the Ar^+ yield with decreasing intensity. This can be regarded as a minor effect, however, as the absolute value of the slope is much smaller at 1,350 nm compared to 800 nm.

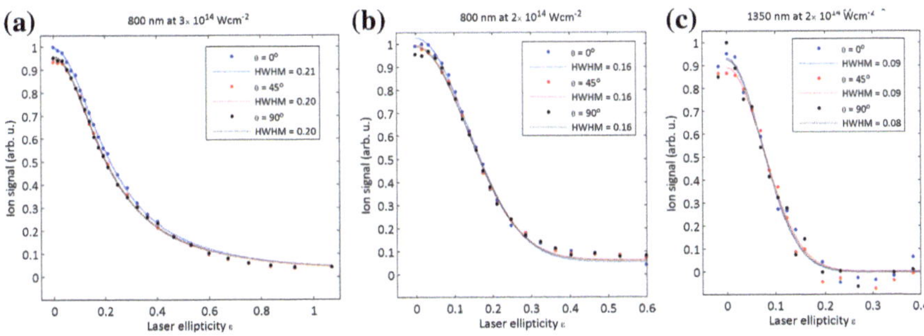

Fig. 6.8 Ellipticity dependence of Ar^{++} formation for different probe pulse wavelengths and intensities: **a** 800 nm, $I_{800}^{(hi)} = 3 \times 10^{14}\,\mathrm{Wcm^{-2}}$, **b** 800 nm, $I_{800}^{(lo)} = 2 \times 10^{14}\,\mathrm{Wcm^{-2}}$ and **c** 1,350 nm, $I_{1350}^{(lo)} = 2 \times 10^{14}\,\mathrm{Wcm^{-2}}$. In all cases, the half width at half maximum (HWHM) of the curves was calculated via numerical fits to the data. In **a** a Lorentzian was employed whilst a Gaussian fit was used in **b** and **c**. Curves were recorded for three different angles θ between the major axis of the probe polarisation ellipse and the linear alignment laser polarisation

Figure 6.8 displays the Ar^{++} yield as a function of probe laser ellipticity ϵ for different probe pulse wavelengths and intensities: (a) 800 nm, $I_{800}^{(hi)} = 3 \times 10^{14}\,\mathrm{Wcm^{-2}}$, (b) 800 nm, $I_{800}^{(lo)} = 2 \times 10^{14}\,\mathrm{Wcm^{-2}}$ and (c) 1,350 nm, $I_{1350}^{(lo)} = 2 \times 10^{14}\,\mathrm{Wcm^{-2}}$. In every case, three values of θ were probed as pointed out in the previous section. Via Lorentzian (a) and Gaussian (b,c) fits, the half width at half maximum (HWHM) $\Delta\epsilon$ of the curves was extracted.

First of all one should note that in each of the three cases, the curves for different values of θ overlap very well and lead to the same $\Delta\epsilon$ within a range of ± 0.01. In (a) a value of $\Delta\epsilon_{800}^{(hi)} = 0.20 \pm 0.01$ is obtained, whereas at an ellpticity of $\epsilon = 0.3$ a significant fraction of the yield compared to the DC offset is still present. For (b) a significantly lower HWHM of $\Delta\epsilon_{800}^{(lo)} = 0.16 \pm 0.01$ was extracted. This leads to the Ar^{++} rate reaching the level of the DC offset in Fig. 6.8b at about $\epsilon = 0.3$. For higher ellipticities, DI in Ar is thus fully suppressed. In (c), the HWHM takes the lowest value out of all three cases with $\Delta\epsilon_{1350}^{(lo)} = 0.09 \pm 0.01$. This strong ellipticity dependence leads to a suppression of Ar^{++} formation for ellipticities $\epsilon > 0.2$. In summary, one observes that $\Delta\epsilon_{800}^{(hi)} > \Delta\epsilon_{800}^{(lo)} > \Delta\epsilon_{1350}^{(lo)}$.

The polarisation dependence of DI in Ar is displayed in Fig. 6.9, comparing probes with 800 and 1,350 nm at $2 \times 10^{14}\,\mathrm{Wcm^{-2}}$. In both cases, the displayed rate is flat with respect to θ with fluctuations around the mean Ar^{++} yield. For 800 nm the fluctuations are about $\pm 2\,\%$, whilst they are higher for 1,350 nm with about $\pm 4\,\%$.

6.4.2 Discussion

As pointed out in the background section, a flat intensity dependence of the ion yield ratio Ar^{++}/Ar^{+} results from NSDI, whilst SDI causes a strong intensity dependence.

This implies that an inflection point in the Ar^{++} rate that leads to a decreasing slope of the ion yield ratio identifies the peak intensity where NSDI begins to dominate over SDI. This is observed in the 800 nm case for a peak intensity of approximately $4 \times 10^{14}\,\mathrm{Wcm}^{-2}$. At the peak intensity associated with $I_{800}^{(lo)} = 2 \times 10^{14}\,\mathrm{Wcm}^{-2}$ the ion yield ratio is mostly independent of intensity and NSDI can thus be regarded as the dominant DI mechanism. Note that the knee structure in the Ar^{++} rate is then directly due to an enhancement of DI caused by NSDI. The above interpretations for the 800 nm case agree extremely well with previously reported measurements [22], which are displayed in Fig. 2.16a. This shows that the experimental setup produces consistent results, which enables comparing the 800 nm case to the much less characterised scenario with 1,350 nm. Here, no inflection point is observed in the Ar^{++} rate, yet the intensity dependence of the ion yield ratio was found to be very weak over the entire peak intensity range from about $3 \times 10^{14}\,\mathrm{Wcm}^{-2}$ to $7 \times 10^{14}\,\mathrm{Wcm}^{-2}$. This implies that for 1,350 nm, NSDI dominates the Ar^{++} formation over the entire intensity range. As was discussed previously, for low intensities this is expected as the longer wavelength leads to higher recollision energies and a smaller Keldysh parameter. It is surprising, however, that NSDI still dominates at much higher peak intensities compared to the 800 nm case. This implies that the onset of SDI as the dominating DI mechanism strongly depends on the driving laser wavelength, such that increasing the wavelength increases the associated peak intensity threshold. This also means that depending on the target species, scenarios may be possible where inelastic recollision processes dominate at very high recollision energies above 100 eV.

It should be noted that the slight negative slope (when going from low to high intensities) observed for 1,350 nm is neither a signature of NSDI (zero slope) nor of SDI (positive slope). The small negative slope thus has to be regarded as an artefact of the measurement. It is most likely caused by a variation of the DC offset of the Ar^{++} peak as a function of intensity that is not fully compensated by the background subtraction procedure in the data analysis. Taking into account the following interpretation of the ellipticity measurements, however, one can argue that this is a minor effect that cannot be the cause of the weak intensity dependence of the ion yield ratio.

Due to focal volume averaging, the intensity dependence of the ion yield ratio is not suitable to unambiguously determine if NSDI dominates the DI mechanism. Here, ellipticity measurements are more adequate as they can be compared to the ellipticity dependence of HHG which is exclusively driven by recollision. In the 800 nm case, the ellipticity curves of two peak intensities (dotted vertical lines in the left panel of Fig. 6.7) can be compared. At the lower intensity $I_{800}^{(lo)} = 2 \times 10^{14}\,\mathrm{Wcm}^{-2}$ the strong ellipticity dependence of HHG is displayed, such that the Ar^{++} rate vanishes for $\epsilon > 0.3$. This implies that no SDI contribution is present at the employed probe laser intensity. Whilst this confirms the interpretation of the recorded intensity curves, the obtained HWHM of $\Delta\epsilon_{800}^{(lo)} = 0.16 \pm 0.01$ also establishes a quantitative threshold for NSDI becoming the exclusive DI mechanism. One may thus conclude that $\Delta\epsilon < 0.16$ implies that NSDI dominates and SDI is fully suppressed, whilst for $\Delta\epsilon > 0.16$ SDI contributions are expected to be present.

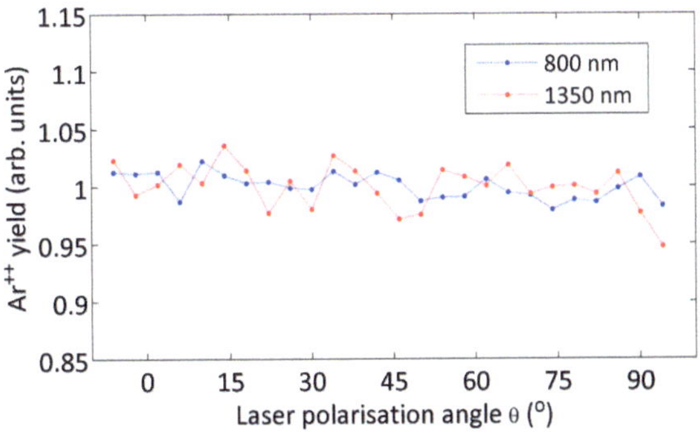

Fig. 6.9 Ar^{++} rate as a function of probe laser polarisation with respect to the alignment laser polarisation. Two probe laser settings were employed: 800 nm, $I_{800}^{(lo)} = 2 \times 10^{14}$ Wcm^{-2} and 1,350 nm, $I_{1350}^{(lo)} = 2 \times 10^{14}$ Wcm^{-2}

In the case of $I_{800}^{(hi)} = 3 \times 10^{14}$ Wcm^{-2}, SDI thus contributes to the Ar^{++} rate, because $\Delta\epsilon_{800}^{(hi)} = 0.20 \pm 0.01$. This also illustrates the limitations of the intensity dependence as an indicator for NSDI. The narrow HWHM and low Ar^{++} yield at $\epsilon = 0.3$ suggest that NSDI still dominates over SDI. The ion yield ratio at the corresponding peak intensity, however, displays a large positive slope. This implies that in this experimental setup, NSDI only becomes visible in the intensity dependence of Ar^{++}/Ar^{+} if SDI is only a negligible contribution to the DI yield.

For $I_{1350}^{(lo)} = 2 \times 10^{14}$ Wcm^{-2}, the HWHM is strongly decreased below the threshold for complete SDI suppression, such that $\Delta\epsilon_{1350}^{(lo)} < \Delta\epsilon_{800}^{(lo)}$. This agrees with the classical picture of the recolliding electron. Due to the longer driving laser wavelength, the recollision time increases. This implies that for the same ellipticity and thus transverse momentum, the electron is driven further away from the parent ion compared to a shorter wavelength. This means that the electron misses the ion at a smaller ellipticity which decreases the HWHM of the associated ellipticity curve. The strong ellipticity dependence also shows that the negative slope of the ion yield ratio present at the associated peak intensity is a minor artefact. This is because the change of the electric field amplitude when increasing its ellipticity to ϵ is negligible, such that the background signal remains constant in this range. Any change in the Ar^{++} yield is thus only due to the ellipticity variation. The ellipticity curve identifies the complete suppression of SDI despite the display of a negative slope at the same intensity. One may thus conclude that the decrease of the slope below zero is an independent effect and the interpretation of the intensity scans is correct.

Figure 6.9 shows that the NSDI yield is independent of θ as expected due to the spherical symmetry of the atom. This shows that there are no artefacts in the polarisation scans that could be caused by distorted polarisation states or reduced collection efficiencies for certain ranges of θ. Furthermore, the measurement establishes a maximum error for the NSDI measurements in CO_2. As the count rates of both species are comparable, the fluctuations of the Ar^{++} yield indicate a maximum error per data point of $\pm 2\,\%$ for 800 nm and $\pm 4\,\%$ at 1,350 nm. The larger fluctuations in the case of 1,350 nm are due to the reduced count rate and increased shot to shot intensity fluctuations of the NIR OPA.

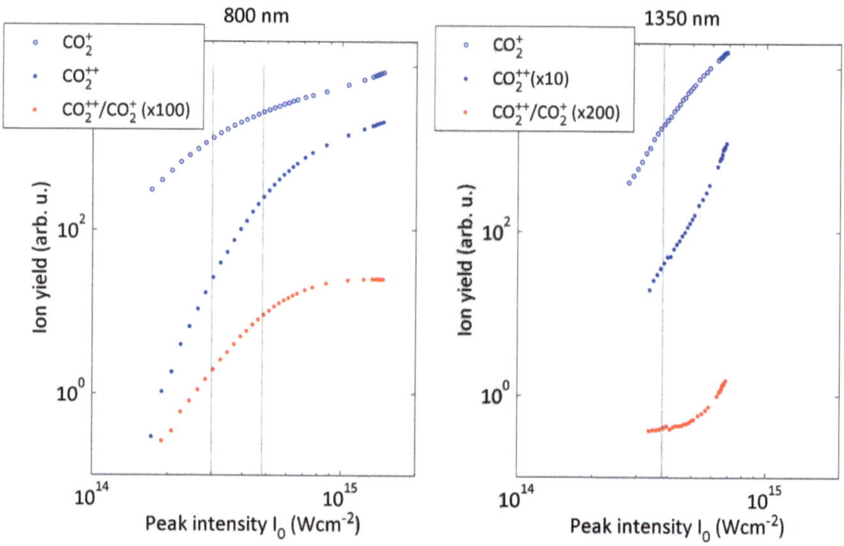

Fig. 6.10 CO_2^+, CO_2^{++} and their ratio CO_2^{++}/CO_2^+ as a function of peak intensity I_0 for 30 fs probe pulses with a wavelength of 800 nm (*left panel*) and 1,350 nm (*right panel*). The *dotted vertical lines* denote the peak intensities employed for conducting ellipticity and polarisation measurements

6.5 Intensity Dependence of NSDI in CO_2

From the same data used for extracting the intensity dependence of DI in Ar, the DI yield of CO_2 was obtained. This enables a direct comparison between Ar and CO_2 under the same experimental conditions.

6.5.1 Results

Figure 6.10 displays the intensity dependence of SI and DI rates in CO_2 and their ratio CO_2^{++}/CO_2^+ as a function of peak intensity I_0 for an 800 nm (left panel) and a 1,350 nm probe pulse (right panel). For 800 nm the CO_2^{++} rate decreases monotonically with decreasing intensity. At high peak intensities above 7×10^{14} Wcm^{-2}, the CO_2^{++} rate displays the same intensity dependence as CO_2^+. Below this intensity, the CO_2^{++} rate decreases faster than CO_2^+. This is reflected in the ion yield ratio CO_2^{++}/CO_2^+. Above 7×10^{14} Wcm^{-2} the intensity dependence is weak and the slope vanishes. Below this intensity the ratio decreases strongly with decreasing intensity and displays an intensity dependence similar to CO_2^+.

For 1,350 nm, however, the CO_2^{++} rate displays an inflection point at approximately 6×10^{14} Wcm^{-2}. Below this peak intensity, the DI yield is enhanced and thus forms a knee structure as in the case of DI of Ar at 800 nm. In the corresponding intensity region, the the CO_2^{++} rate takes the same intensity dependence as CO_2^+, which results in a vanishing slope of their ratio CO_2^{++}/CO_2^+.

6.5.2 Discussion

At 800 nm, the large slope of the intensity dependence of CO_2^{++}/CO_2^{+} shows that NSDI never becomes the dominant DI mechanism over the whole intensity range. For peak intensities above $7 \times 10^{14}\,\mathrm{Wcm^{-2}}$ the vanishing slope cannot be due to a suppression of SDI, as sequential ionisation is favoured for increasing intensities. This effect must thus be attributed to ionisation saturation of CO_2^{++} and CO_2^{+}. Here, all CO_2 molecules in the interaction region are ionised such that an increase in intensity causes a change in ionisation rate only via a geometrical change of the spatial intensity distribution at the focus. This focal volume effect is the same for all saturated ionisation channels leading to the same intensity dependence above the saturation threshold. The results for 800 nm agree very well with previously reported results, where a flat intensity dependence of CO_2^{++}/CO_2^{+} was found for intensities below $2 \times 10^{14}\,\mathrm{Wcm^{-2}}$ [24]. The corresponding data is displayed in Fig. 2.16b.

At 1,350 nm, however, NSDI becomes the dominant DI mechanism for peak intensities below $6 \times 10^{14}\,\mathrm{Wcm^{-2}}$. In this intensity region the slope of the intensity dependence of CO_2^{++}/CO_2^{+} vanishes and one may thus expect complete suppression of SDI in CO_2 at $I_{1350}^{(lo)} = 2 \times 10^{14}\,\mathrm{Wcm^{-2}}$ (dotted vertical line). Taking into account the results in 800 nm reported in [24], this means that the intensity threshold for the onset of SDI in CO_2 is shifted from about $2 \times 10^{14}\,\mathrm{Wcm^{-2}}$ to $6 \times 10^{14}\,\mathrm{Wcm^{-2}}$. This is an interesting result that enables isolating recollision induced double ionisation whilst opening all possible channels in the underlying mechanism due to the high recollision energies acquired in the long wavelength, high intensity probe laser field.

6.6 Ellipticity Dependence of NSDI in CO_2

The ellipticity dependence of DI in CO_2 at $2 \times 10^{14}\,\mathrm{Wcm^{-2}}$ can be used to check the conclusions from the intensity measurements. With SDI fully suppressed, ellipticity curves at different orientations θ of the molecular alignment distribution can then be used to map out the shape of the recolliding electron wavepacket in momentum space.

6.6.1 Results

Figure 6.11 displays the ellipticity dependence of DI in impulsively aligned CO_2, where θ denotes the angle of the major axis of the probe polarisation ellipse with respect to the molecular axis. In part (a), the ellipticity curves from 800 nm and 1,350 nm probe pulses at $2 \times 10^{14}\,\mathrm{Wcm^{-2}}$ and $\theta = 45°$ are compared. This angle was chosen as it lead to the highest yield of CO_2^{++}. In addition, the ellipticity curve for Ar^{++} at 1,350 nm and the same intensity is displayed as a reference. Part (b) compares the ellipticity dependence of CO_2^{++} formation for different values of θ obtained

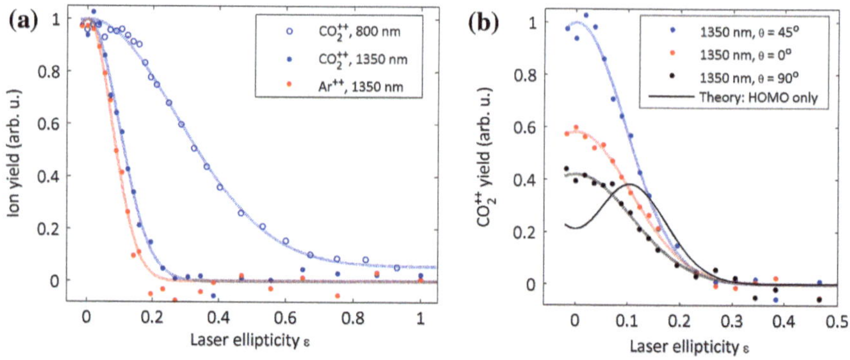

Fig. 6.11 **a** Comparison of the ellipticity dependence of CO_2^{++} for 800 and 1,350 nm probe laser pulses at 2×10^{14} Wcm^{-2} and Ar^{++} at 1,350 nm and 2×10^{14} Wcm^{-2}. All curves were obtained for $\theta = 45°$, where θ is the angle of the major axis of the probe polarisation ellipse with respect to the molecular axis. **b** CO_2^{++} rate as a function of ellipticity ϵ for 1,350 nm probe laser pulses at 2×10^{14} Wcm^{-2} and different values of θ. The theoretical curve is obtained from an analytical model and corresponds to $\theta = 90°$ (Part **(b)** reprinted with permission from [32]. Copyright (2013) by the American Physical Society)

via 1,350 nm probe pulses at 2×10^{14} Wcm^{-2}. Additionaly, a theoretical curve is displayed, which is associated with $\theta = 90°$ and was calculated via an analytical model, which is presented in Sect. 6.2. It estimates the shape of the ellipticity curve under the employed experimental conditions with the DI mechanism restricted to the highest occupied molecular orbital (HOMO).

To all experimental data curves, a Gaussian was fitted and its HWHM $\Delta\epsilon$ extracted. This allowed for a quantitative comparison between the strength of the ellipticity dependence of CO_2^{++} for 800 and 1,350 nm. In the case of a 800 nm probe, $\Delta\epsilon_{800} = 0.29 \pm 0.02$ was extracted for all values of θ (only the data for $\theta = 45°$ is displayed in Fig. 6.11a). Note that at $\epsilon = 0.3$ the CO_2^{++} yield is still about 50 % compared to the yield at linear polarisation. In comparison, the curve at 1,350 nm displays a significantly stronger ellipticity dependence. Here, a value of $\Delta\epsilon_{1350} = 0.12 \pm 0.02$ was extracted via Gaussian fits and was the same for all values of θ within its error range. Additionally, the CO_2^{++} rate vanishes for $\epsilon > 0.3$ at 1,350 nm. The curve for Ar^{++} displays a very similar behaviour, albeit with a slightly stronger ellipticity dependence. This manifests itself in a smaller HWHM of $\Delta\epsilon_{1350}^{(lo)} = 0.09 \pm 0.01$ (see previous section).

Figure 6.11b allows for a comparison of the ellipticity dependence of the CO_2^{++} rate in terms of the orientation of the molecule with respect to the laser polarisation ellipse. Here, all measured ellipticity curves reach their maximum at $\epsilon = 0$, where the probe laser pulse is linearly polarised. There, the relative differences in the CO_2^{++} yield for different values of θ already indicate its polarisation dependence. This is discussed in detail in Sect. 6.7. The shape of all ellipticity curves is very well described by a Gaussian dependence as shown by the associated numerical fits. This is in contrast with the theoretical curve that estimates the ellipticity dependence of

NSDI from HOMO at $\theta = 90°$ under the employed experimental conditions. It was obtained from the analytical model outlined in Sect. 6.2. The curve diplays a strong suppression of the CO_2^{++} rate for $\epsilon = 0$ and reaches its maximum at approximately $\epsilon = 0.13$.

6.6.2 Discussion

The first major finding of the ellipticity measurements is that at 2×10^{14} Wcm^{-2} and 1,350 nm CO_2^{++} formation is exclusively driven by NSDI, whilst at 800 nm a substantial contribution of SDI is present. This is because $\Delta\epsilon_{1350} = 0.12 \pm 0.02 < 0.16$, whilst $\Delta\epsilon_{800} = 0.29 \pm 0.02 > 0.16$, following the conclusions from the ellipticity measuements in Ar^{++}. This agrees with the intensity dependence of CO_2^{++} at 800 and 1,350 nm. Hence at 1,350 nm, the recollision event can be characterised via ellipticity and polarisation measurements as direct laser induced double ionisation is fully suppressed. This is further supported by the observation that at 1,350 nm and 2×10^{14} Wcm^{-2}, the CO_2^{++} rate is as strongly dependent on the laser ellipticity as the Ar^{++} rate. Here, the slightly reduced value for the HWHM $\Delta\epsilon$ in Ar is most probably due to the smaller size of the atom compared to the CO_2 molecule.

This enables the second major finding of the ellipticity curves obtained with 1,350 nm for different values of θ. As the CO_2^{++} formation is purely driven by recollision, the ellipticity curves map out the shape of the recolliding electron wavepacket in momentum space. As was pointed out in the background section, the molecular structure of the ionised orbital is encoded in the wavepacket's shape. In particular, if ionisation takes place along a nodal plane of the ionised orbital, the ellipticity curve will possess a dip at $\epsilon = 0$ and reach a maximum at nonzero ellipticity. One may thus expect that if NSDI is dominated by tunneling ionisation from the HOMO, the ellipticity curves at $\theta = 0°$ and $\theta = 90°$ should display a dip at $\epsilon = 0$. However, the visibility of this effect in an experiment depends on the width of the nodal plane and the degree of molecular alignment. This was discussed in detail in Sect. 6.2, where an analytical model was developed that estimates the visibility of a dip under the employed experimental conditions if only the HOMO is taken into account. The results showed that for the degree of alignment present in this experiment, the nodal plane at $\theta = 0°$ can most probably not be resolved in an ellipticity measurement due to the narrow node. At $\theta = 90°$, however, the node is significantly wider and the suppression of NSDI at $\epsilon = 0$ is expected to be visible as a pronounced dip in the associated ellipticity curve. This is shown by the theoretical curve in Fig. 6.11. Yet, the corresponding experimental curve (black data points in the same figure) does not display this behaviour. Instead, the maximum NSDI yield is reached for $\epsilon = 0$ and the shape of the recolliding electron wavepacket is described by a Gaussian dependence.

By comparing the results of the analytical model to the experimental data, one may now examine the possible reasons for the absence of the ellipticity dip at $\theta = 90°$. The visibility of the dip in the theoretical curve relies on the width of the nodal plane and the degree of alignment in the molecular ensemble. Both issues were taken into account in the model and it was shown in Sect. 6.2 that even for a very narrow nodal plane, the degree of alignment in the experiment is expected to be sufficient to lead to a visible ellipticity dip at $\theta = 90°$ (see Fig. 6.4). However, the model only takes into account a single tunneling-ionisation channel via HOMO. It may thus be possible that in the experiment additional channels are contributing involving ionisation from lower lying molecular orbitals that do not possess a nodal plane at $\theta = 90°$. In this case, the CO_2^{++} yield would not be fully suppressed due to the presence of the additional channels and the dip in the ellipticity dependence could not be observed.

This explanation for the absence of the ellipticity dip relies on the assumption that the analytical model leads to an appropriate estimate of its visibility in an experiment. This can only be tested unambiguously when employing the model in a scenario where an ellipticity dip is resolved experimentally. However, it was not possible to conduct such a test in the used experimental setup, due to the lack of a suitable target molecule. In particular, one requires a molecule where the initial TI step proceeds via a single orbital that has a sufficiently wide nodal plane. As such a molecule was not available, the presence of multiple NSDI channels was investigated via an additional measurement. In particular, the polarisation dependence of NSDI in CO_2 was studied, providing an independent test for the involvement of additional channels that proceed via orbitals other than the HOMO.

6.7 Polarisation Dependence of NSDI in CO_2

In order to identify the specific molecular states participating in the NSDI mechanism, the polarisation dependence of CO_2^+ and CO_2^{++} was recorded in the molecular frame. In both cases, the angularly resolved rates encode the orbital structure of the molecular states participating in the NSDI mechanism. As for the ellipticity measurement, 1,350 nm, 30 fs probe pulses were employed at $2 \times 10^{14}\,\text{Wcm}^{-2}$, to ensure that contributions from SDI are fully suppressed.

6.7.1 Results

Figure 6.12 displays the angular SI rate (CO_2^+) and the NSDI rate (CO_2^{++}) as a function of θ with $\epsilon = 0$. Here, the total SI rate is the sum of all ion yields that result from the parent ion, which includes fragments from dissociative single ionisation channels. Each data set is fitted with a polynomial and normalised with respect to the maximum of the associated fit.

The SI rate is suppressed at $\theta = 0°$ and $\theta = 90°$, whilst this effect is much stronger for perpendicular polarisation. The maximum SI rate is reached at $\theta_{max}^{SI} = 39° \pm 1°$, with the error estimated from the variation of the fit with polynomial order. The angularly resolved NSDI rate displays significant differences compared to the SI rate. Whilst CO_2^{++} formation is also suppressed at $\theta = 0°$ and $\theta = 90°$, this effect is significantly stronger compared to the suppression of CO_2^+ formation at these angles.

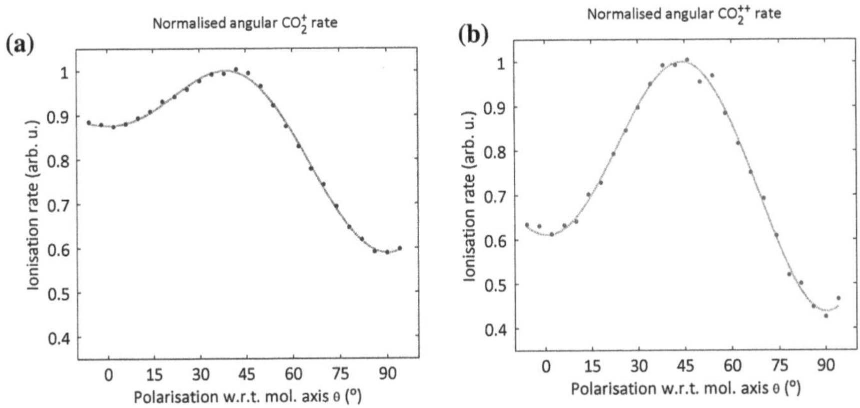

Fig. 6.12 CO_2^+ (**a**) and CO_2^{++} (**b**) as a function of angle θ between the linear probe laser polarisation and the molecular axis. 1,350 nm, 30 fs probe pulses at $2 \times 10^{14}\,\mathrm{Wcm}^{-2}$ were employed for the measurement

Furthermore, the difference between the CO_2^{++} rates at parallel and perpendicular polarisation is reduced compared to the corresponding CO_2^+ rates. The angular NSDI rate is maximised at $\theta_{max}^{NSDI} = 44° \pm 1°$.

6.7.2 Discussion

As was pointed out in the background section, the electronic density profiles of the tunnel ionised orbitals are encoded in the angular TI rate. Due to the pronounced suppression of the CO_2^+ formation for parallel and perpendicular polarisation, the obtained angular rate suggests that in this experiment, SI of CO_2 is strongly dominated by the HOMO. Without additional information, however, the angular SI rate does not allow for the reconstruction of the branching ratio of this ionisation channel in the total CO_2^+ yield. Yet, the extracted position of the maximum rate at $\theta_{max}^{SI} = 39° \pm 1°$ agrees extremely well with recent theoretical calculations of the angularly resolved, direct ionisation rate from the HOMO of CO_2, where the maximum rate was found at approximately 40° [17]. The corresponding curve is displayed in Fig. 2.15a. In fact, due to its π_u symmetry, contributions from the next lower lying orbital HOMO-1 would enhance the SI rate most strongly for angles $\theta > 40°$. This would shift the position of the maximum CO_2^+ yield towards larger angles, such that $\theta_{max}^{SI} > 40°$. One may thus conclude that under the employed experimental conditions, CO_2^+ formation almost exclusively proceeds via TI of the HOMO with significantly lower contributions from lower lying orbitals.

The angular NSDI rate is proportional to an angular cross section $\sigma^{NSDI}(\theta)$ that must be a convolution of the angular SI cross section $\sigma^{SI}(\theta)$ and the electron-ion recollision cross section $\sigma^{r}(\theta)$. The experimental results show that $\theta_{max}^{NSDI} > \theta_{max}^{SI}$, which implies that $\sigma^{NSDI}(\theta)$ differs significantly from $\sigma^{SI}(\theta)$. This suggests that the

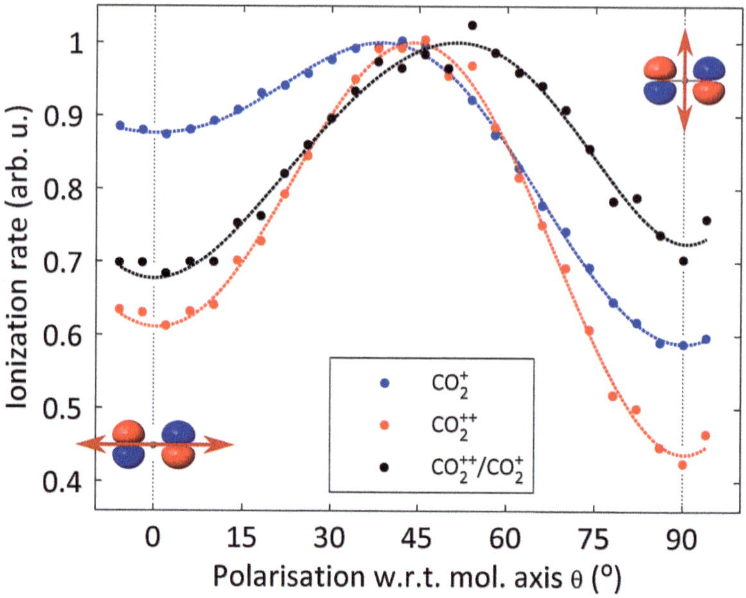

Fig. 6.13 Ion yield ratio CO_2^{++}/CO_2^+ as a function of angle θ between the linear probe laser polarisation and the molecular axis. 1,350 nm, 30 fs probe pulses at 2×10^{14} Wcm^{-2} were employed for the measurement. The angular single and double ionisation rates are included for comparison

recollision cross section $\sigma^r(\theta)$ does not scale like the electronic density profile of the HOMO. In fact, the relative angular recollision cross section can be extracted from the data via the following arguments. The ellipticity measurements presented in Sect. 6.6 show that the shape of the recolliding electron wavepacket is independent of θ. This implies that the propagation of the free electron in the continuum has no effect on the angular NSDI cross section, which is thus solely determined by the SI and recollision cross section. In this case one obtains $\sigma^{NSDI}(\theta) \propto \sigma^{SI}(\theta) \times \sigma^r(\theta)$, such that the ion yield ratio CO_2^{++}/CO_2^+ is a direct measure of $\sigma^r(\theta)$. The ratio between the angular NSDI and SI rates was therefore directly calculated from the corresponding data displayed in Fig. 6.12 and is displayed in Fig. 6.13. One observes that the ratio is suppressed at parallel and perpendicular polarisation. Compared to the SI rate, the ratio is suppressed more strongly at $\theta = 0°$, whilst at $\theta = 90°$, the suppression is less pronounced than for SI. The maximum of the ratio curve was found at $\theta^r_{max} = 52° \pm 1°$.

The comparison of $\sigma^{SI}(\theta)$ and the extracted $\sigma^r(\theta)$ then allows for the characterisation of the recollision event in terms of the orbitals involved in the electron-ion collision. This can be seen as follows. During the inelastic recollision event, the free electron interacts with the bound electrons of the parent ion. Hence, one may infer that the collision cross section scales like the electron density profile of the orbital of the molecular state of the parent ion. This approximation is valid for sufficiently high collision energies. Due to the long driving laser wavelength of 1,350 nm very high maximum recollision energies were achieved in this experiment. At $2 \times 10^{14}\,\mathrm{Wcm}^{-2}$ this resulted in $E^r_{1350} \approx 109\,\mathrm{eV}$ compared to $E^r_{800} \approx 38\,\mathrm{eV}$ at 800 nm. Via crossed beam experiments, the maximum cross section for electron-CO_2^+ collisions has been found at an electron energy of about 130 eV [33]. As the recollision energies present in this experiment are of the same order, one may thus conclude that the above approximation is valid. Furthermore, the electronic density profile of the involved ionic states are obtained from the Koopmans approximation (see Sect. 2.3.2). Here, this implies that TI from HOMO results in a parent ion in the ground state X with the same orbital symmetry. In the same manner, HOMO-1 is associated with the first excited ionic state A.

With the above approximations, the angular dependence of the recollision cross section may now be analysed in terms of the involved ionic states in the same way as the angular SI rate. Firstly this means that the suppression of $\sigma^r(\theta)$ at $\theta = 0°$ and $\theta = 90°$ indicate a strong contribution of the ionic ground state X. However, the angular position of the maximum cross section fulfills $\theta^r_{max} > 40°$. The NSDI mechanism can thus not be restricted to a single channel via the HOMO and the ionic ground state of CO_2. Instead, the shift of the maximum angular cross section to larger angles requires the contribution of ionic states that do not possess a nodal plane perpendicular to the molecular axis. Of the next lower lying orbitals only the excited ionic states A and C match this criterion. It is well known, however, that the third excited state C of CO_2^+ fully predissociates [34], such that its involvement in the NSDI mechanism can be excluded. One may thus conclude that the polarisation dependence of NSDI in CO_2 identifies multiple channels in the underlying mechanism that proceed via the ground and first excited ionic state.

This conclusion is fully consistent with the results from the ellipticity dependence of NSDI in CO_2, where the absence of a dip at $\epsilon = 0$ for $\theta = 90°$ also required the contribution of an NSDI channel involving a molecular state without a node at this angle. One should note, however, that even though HOMO and HOMO-1 both have a nodal plane at $\theta = 0°$, it is unlikely that the resulting dip at $\epsilon = 0$ can be resolved in this experiment. This is because of the small width of the nodal planes and an insufficient degree of molecular alignment, as was pointed out in Sect. 6.2.

6.7.3 Multichannel NSDI

The ellipticity and polarisation measurements presented in this chapter identified the contribution of the HOMO and HOMO-1 and their associated ionic states X and A to the NSDI mechanism in CO_2. This implies that the underlying double ionisation mechanism can consist of three possible channels illustrated in Fig. 6.14: (i) TI from HOMO followed by impact ionization of the ionic X state, (ii) TI from HOMO-1 followed by impact ionization of the ionic A state, (iii) TI from HOMO followed by recollision excitation with subsequent field ionization (RESI) [19] via the ionic A state. Due to the high maximum recollision energy of $E^r_{1350} \approx 109\,\mathrm{eV}$, all channels are open. As the ellipticity dependence of NSDI encodes the nodal plane of the

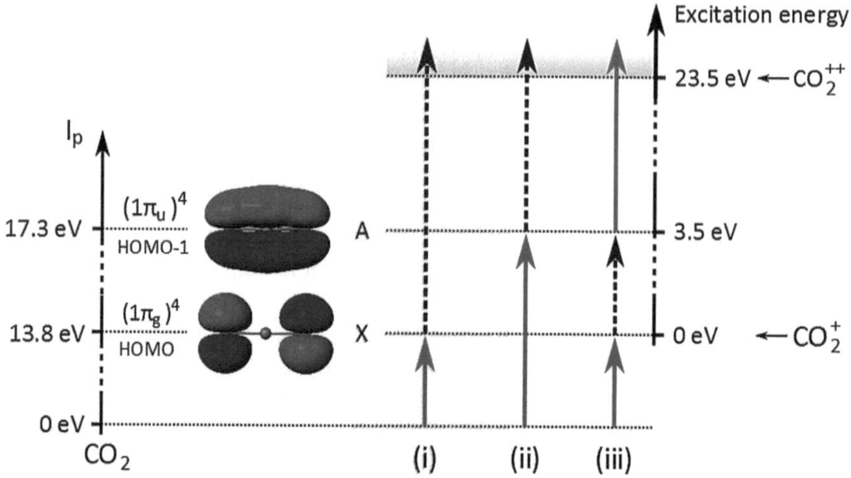

Fig. 6.14 Schematic illustration of the three possible channels contributing to NSDI in CO_2. *Solid arrows* denote tunneling ionisation and *dashed arrows* inelastic recollision. The second ionisation continuum is reached at 23.5 eV with respect to the ionic ground state [35] (Reprinted with permission from [32]. Copyright (2013) by the American Physical Society)

tunnel ionised orbital, the absence of the dip at $\theta = 90°$ requires TI from HOMO-1. This means that channel (ii) is strongly pronounced in the NSDI mechanism. The suppression of $\sigma^r(\theta)$ at $\theta = 90°$ requires channels involving recollision with the ionic ground state X. This is the case for channel (i) and (iii). However, it is not possible to distinguish the two channels on the basis of the obtained experimental results. Furthermore, it is not possible to extract the individual branching ratios of the contributing channels.

One may now ask why channel (ii) plays such a pronounced role in the NSDI mechanism. First of all it should be noted that the angular CO_2^+ rate suggests that TI in this experiment mostly proceeds via the HOMO. However, this result does not exclude TI from HOMO-1. It only shows that its rate is significantly less than the TI rate from HOMO. The remaining results from the polarisation and ellipticity measurements then show that the relatively small fraction of TI from HOMO-1 present in the experiment leads to a very large NSDI contribution at $\theta = 90°$. Here, channel (ii) has a comparable overall probability compared to (i) and (iii). The reason for this is that at this angle, most recolliding electrons from channels (i) and (iii) miss the parent ion due to their non-zero transverse momentum.

It should also be noted that the first excited ionic state A possesses a bonding orbital, such that the removal of an electron from this state would lead to a CO_2^{++} ion in a dissociative state. The reason why channels (ii) and (iii) are still expected to contribute to the detected NSDI yield is that within the Koopmans approximation, both channels lead to preferential formation of CO_2^{++} ions in the first excited state. It has been found that this state is part of a band of metastable states that display a life time that is sufficiently long for the ions to be detected in this experiment [35].

6.8 Conclusion

In the experiments presented in this chapter, multichannel contributions to molecular NSDI are identified for the first time. For achieving this, a long wavelength drive field was used to ensure that NSDI is the only route to CO_2^{++}. Employing a TOF spectrometer and impulsive molecular alignment, the inelastic rescattering event was fully characterised by measuring the shape of the returning electron wavepacket and the recollision cross section in the molecular frame. The results provide strong evidence for the contribution of the ionic ground and first excited state to the NSDI mechanism due to their distinct orbital symmetry. The results also provide the first angularly resolved electron-ion collision measurement in CO_2, albeit over a finite energy range. The universal applicability of this method thus demonstrates the potential of using inelastic recollisions in the molecular frame of reference for extracting and controlling molecular dynamics.

References

1. X.M. Tong, Z.X. Zhao, C.D. Lin, Theory of molecular tunneling ionization. Phys. Rev. A **66**(3), 033402 (2002)
2. O. Smirnova, Y. Mairesse, S. Patchkovskii, N. Dudovich, D. Villeneuve, P. Corkum, M.Y. Ivanov, High harmonic interferometry of multi-electron dynamics in molecules. Nature **460**(7258), 972–977 (2009)
3. P.B. Corkum, Plasma perspective on strong field multiphoton ionization. Phys. Rev. Lett. **71**, 1994–1997 (1993)
4. S. Baker, J.S. Robinson, C.A. Haworth, H. Teng, R.A. Smith, C.C. Chiril, M. Lein, J.W.G. Tisch, J.P. Marangos, Probing proton dynamics in molecules on an attosecond time scale. Science **312**(5772), 424–427 (2006)
5. H.J. Worner, J.B. Bertrand, D.V. Kartashov, P.B. Corkum, D.M. Villeneuve, Following a chemical reaction using high-harmonic interferometry. Nature **466**(7306), 604–607 (2010)
6. J. Itatani, J. Levesque, D. Zeidler, H. Niikura, H. Pepin, J.C. Kieffer, P.B. Corkum, D. M. Villeneuve, Tomographic imaging of molecular orbitals. Nature **432**(7019), 867–871 (2004)
7. R. Torres, T. Siegel, L. Brugnera, I. Procino, J.G. Underwood, C. Altucci, R. Velotta, E. Springate, C. Froud, I.C.E. Turcu, M.Y. Ivanov, O. Smirnova, J.P. Marangos, Extension of high harmonic spectroscopy in molecules by a 1300 nm laser field. Opt. Express **18**(3), 3174–3180 (2010)
8. L. Keldysh, Ionization in the field of a strong electromagnetic wave. Sov. Phys. JETP **20**, 1307–1314 (1965)
9. B.K. McFarland, J.P. Farrell, P.H. Bucksbaum, M. Guhr, High harmonic generation from multiple orbitals in N_2. Science **322**, 1232–1235 (2008)
10. W. Boutu, S. Haessler, H. Merdji, P. Breger, G. Waters, M. Stankiewicz, L.J. Frasinski, R. Taieb, J. Caillat, A. Maquet, P. Monchicourt, B. Carre, P. Salieres, Coherent control of attosecond emission from aligned molecules. Nat. Phys. **4**(7), 545–549 (2008)
11. X. Jia, W. Li, J. Liu, J. Chen, Alignment-dependent nonsequential double ionization of N_2 in intense laser fields: the role of different valence orbitals. Phys. Rev. A **80**, 053405 (2009)
12. D. Pavicic, K.F. Lee, D.M. Rayner, P.B. Corkum, D.M. Villeneuve, Direct measurement of the angular dependence of ionization for N_2, O_2, and CO_2 in intense laser fields. Phys. Rev. Lett. **98**(24), 243001 (2007)

13. M. Abu-samha, L.B. Madsen, Theory of strong-field ionization of aligned CO_2. Phys. Rev. A **80**(2), 023401 (2009)
14. G.A. Gallup, I.I. Fabrikant, Semiclassical complex-time method for tunneling ionization: molecular suppression and orientational dependence. Phys. Rev. A **81**(3), 033417 (2010)
15. S. Petretti, Y.V. Vanne, A. Saenz, A. Castro, P. Decleva, Alignment-dependent ionization of N_2, O_2, and CO_2 in intense laser fields. Phys. Rev. Lett. **104**(22), 223001 (2010)
16. R. Murray, M. Spanner, S. Patchkovskii, M.Y. Ivanov, Tunnel ionization of molecules and orbital imaging. Phys. Rev. Lett. **106**(17), 173001 (2011)
17. L. Torlina, M. Ivanov, Z.B. Walters, O. Smirnova, Time-dependent analytical r-matrix approach for strong-field dynamics. II. many-electron systems. Phys. Rev. A **86**, 043409 (2012)
18. D.N. Fittinghoff, P.R. Bolton, B. Chang, K.C. Kulander, Polarization dependence of tunneling ionization of helium and neon by 120-fs pulses at 614 nm. Phys. Rev. A **49**, 2174 (1994)
19. B. Feuerstein, R. Moshammer, D. Fischer, A. Dorn, C.D. Schrter, J. Deipenwisch, J.R.C. Lopez-Urrutia, C. Höhr, P. Neumayer, J. Ullrich, H. Rottke, C. Trump, M. Wittmann, G. Korn, W. Sandner, Separation of recollision mechanisms in nonsequential strong field double ionization of Ar: the role of excitation tunneling. Phys. Rev. Lett. **87**, 043003 (2001)
20. A. Rudenko, T. Ergler, K. Zrost, B. Feuerstein, V.L.B. de Jesus, C.D. Schrter, R. Moshammer, J. Ullrich, From non-sequential to sequential strong-field multiple ionization: identification of pure and mixed reaction channels. J. Phys. B: At. Mol. Opt. Phys. **41**(8), 081006 (2008)
21. A.S. Alnaser, D. Comtois, A.T. Hasan, D.M. Villeneuve, J. Kieffer, I.V. Litvinyuk, Strong-field non-sequential double ionization: wavelength dependence of ion momentum distributions for neon and argon. J. Phys. B: At. Mol. Opt. Phys. **41**(3), 031001 (2008)
22. C. Guo, M. Li, J.P. Nibarger, G.N. Gibson, Single and double ionization of diatomic molecules in strong laser fields. Phys. Rev. A **58**, R4271 (1998)
23. C. Cornaggia, P. Hering, Nonsequential double ionization of small molecules induced by a femtosecond laser field. Phys. Rev. A **62**(2), 023403 (2000)
24. L. Pei, C. Guo, Nonsequential double ionization of triatomic molecules in strong laser fields. Phys. Rev. A **82**(2), 021401 (2010)
25. K.S. Budil, P. Salies, M.D. Perry, A. L'Huillier, Influence of ellipticity on harmonic generation. Phys. Rev. A **48**, R3437 (1993)
26. V.R. Bhardwaj, D.M. Rayner, D.M. Villeneuve, P.B. Corkum, Quantum interference in double ionization and fragmentation of C_6H_6 in intense laser fields. Phys. Rev. Lett. **87**, 253003 (2001)
27. A.S. Alnaser, C.M. Maharjan, X.M. Tong, B. Ulrich, P. Ranitovic, B. Shan, Z. Chang, C.D. Lin, C.L. Cocke, I.V. Litvinyuk, Effects of orbital symmetries in dissociative ionization of molecules by few-cycle laser pulses. Phys. Rev. A **71**, 031403 (2005)
28. D. Zeidler, A. Staudte, A.B. Bardon, D.M. Villeneuve, R. Dorner, P.B. Corkum, Controlling attosecond double ionization dynamics via molecular alignment. Phys. Rev. Lett. **95**, 203003 (2005)
29. J. Liu, D.F. Ye, J. Chen, X. Liu, Complex dynamics of correlated electrons in molecular double ionization by an ultrashort intense laser pulse. Phys. Rev. Lett. **99**, 013003 (2007)
30. Y. Li, J. Chen, S.P. Yang, J. Liu, Alignment effect in nonsequential double ionization of diatomic molecules in strong laser fields. Phys. Rev. A **76**(2), 023401 (2007)
31. S.J. Weber, M. Oppermann, M.Y. Ivanov, J.P. Marangos, Ellipticity dependence of strong field electron recollision for molecular structural imaging. J. Mod. Opt. **60**, 1379–1387 (2013)
32. M. Oppermann, S.J. Weber, L.J. Frasinski, M.Y. Ivanov, J.P. Marangos, Multichannel contributions in the nonsequential double ionization of CO_2. Phys. Rev. A **88**, 043432 (2013)
33. E.M. Bahati, J.J. Jureta, D.S. Belic, S. Rachafi, P. Defrance, Electron impact ionization and dissociation of CO_2^+ to C^+ and O^+. J. Phys. B: At. Mol. Opt. Phys. **34**, 1757–1767 (2001)
34. R. Bombach, J. Dannacher, J.-P. Stadelmann, J.C. Lorquet, Branching ratios and partition of the excess energy for the predissociation of CO_2^+ $\tilde{C}^2\Sigma_g^+$ molecular cations. J. Chem. Phys. **79**, 4214 (1983)
35. A.E. Slattery, T.A. Field, M. Ahmad, R.I. Hall, J. Lambourne, F. Penent, P. Lablanquie, J.H.D. Eland, Spectroscopy and metastability of CO_2^{2+} molecular ions, J. Chem. Phys. **122**, 084317–084317-13 (2005)

Chapter 7
Strong Field Control of Dissociative Excitation in CO_2^+

In this chapter the molecular structure dependence of strong field induced dissociation in CO_2 is investigated. In particular, the dissociative excitation of CO_2^+ was studied in the molecular frame as a function of probe laser intensity, ellipticity and polarisation direction with respect to the molecular bond. This enabled the identification of the main excitation pathway consisting of tunneling ionisation from $HOMO^{-2}$ followed by a parallel dipole transition from the second excited state B to the predissociating state $CO_2^+(C^2\Sigma_g^+)$, whilst recollision excitation was shown to not play a role. Using laser induced impulsive alignment, the strong field induced coupling of the ionic states B and C could thus be controlled by the laser polarisation. This lead to a suppression of the fragmentation yield of up to 70 % when comparing the yields from perpendicular and parallel polarisation with respect to the molecular axis. It was then found that the dissociation mechanism preferentially produces $CO^+_{(1,0)}$ ions with a very large branching ratio for this channel between $R_b = 0.8$ and 0.9. This implies that the molecular parent ion was mostly prepared in vibrationally excited states and remained in a linear configuration during the predissciation mechanism.

7.1 Background

Photo-induced dissociation is of special interest in molecular physics. The reason for this is that it always involves excitation processes, such that the associated pathways and molecular state populations may be studied as a function of laser pulse parameters. Furthermore, the application of femtosecond laser pulses allows for the time resolution of these dynamics on the time scale of the nuclear motion [1]. This perspective has informed the application of ultrafast strong field processes for probing molecular dynamics, with the aim to extend the study of molecular dynamics to the investigation of the electronic degrees of freedom and their coupling to the nuclear degrees of freedom. Recent examples for achieving this are the use of high harmonic generation [2], time-resolved photoelectron spectroscopy [3] and strong

M. Oppermann, *Resolving Strong Field Dynamics in Cation States of CO_2 via Optimised Molecular Alignment*, Springer Theses, DOI: 10.1007/978-3-319-05338-7_7,

field ionisation [4], to name but a few highlights from the last 2 years. In addition, the ongoing development of laser sources providing attosecond pulses promises tools for directly resolving molecular dynamics on the electronic time scale in a weak field regime [5].

Against this backdrop of research, the study of strong field induced dynamics in CO_2 is extended to dissociation in this chapter. In particular, the dissociative excitation mechanism (DE) of CO_2^+ is investigated. Here, DE may either lead to CO^+ or O^+ ions hereafter referred to as the $CO^+_{(1,0)}$ and $O^+_{(1,0)}$ channels respectively, where the subscripts denote the distribution of the available ionic charge. The aim of the experiment is to identify and control the relevant strong field induced excitation pathways by studying the corresponding fragmentation yields as a function of driving laser parameters such as intensity, ellipticity and polarisation in the molecular frame of reference.

7.1.1 Predissociation Mechanism

It has been shown that the lowest three states of CO_2^+ - X, A, B—are mostly stable within at least 1 μs [8]. The reason being that their associated dissociation continua cannot be reached via a vertical Franck-Condon transition from the CO_2 ground state [9]. The third excited ionic state C is the first state above the lowest dissociation limit. This is displayed in Fig. 7.1a, where the relevant potential energy curves of CO_2^+ and its lowest dissociation limits are shown. It has been found that the C state fully predissociates and dominates the fragmentation mechanism. Several theoretical and experimental studies have thus been performed to understand the corresponding dissociation dynamics [6, 7, 9–13]. Due to the dissociation limits of the respective fragments, $O^+_{(1,0)}$ may be produced from its vibrational ground state, whereas $CO^+_{(1,0)}$ can only result from vibrationally excited states. The theoretical work of Praet and coworkers [9] has established the basic predissociation mechanism and has later been extended only slightly [7, 12, 13]. As shown in Fig. 7.1a, the lowest dissociation limit resulting in $CO^+(X^2\Sigma^+)$ and $O^+(^4S_u)$ are accessed via the repulsive states $b^4\Pi_u^+$ and $a^4\Sigma_g^-$ respectively. The associated pathways depend on the vibrational excitation of the C state and the bending mode of the molecule. When bent, the $b^4\Pi_u^+$ state is denoted as $^4B_1$. When prepared in the vibrational ground state (0, 0, 0), the only available dissociation pathway is given by [13]:

$$CO_2^+(C^2\Sigma_g^+; 0,0,0) \rightarrow CO_2^+(A^2\Pi_u) \rightarrow CO_2^+(a^4\Sigma_g^-) \rightarrow O^+(^4S_u) + CO(X^1\Sigma^+) \tag{7.1}$$

In this pathway, the first transition is established via vibronic and the second via spin-orbit coupling. For higher vibrational states, dissociation to $CO^+(X^2\Sigma^+)$ becomes possible via the state $b^4\Pi_u^+$. Its spin-orbit coupling with the vibrationally excited $CO_2^+(C^2\Sigma_g^+)$ state is particularly strong such that population transfer is very efficient. The $b^4\Pi_u^+$ state may then directly dissociate via the following pathway:

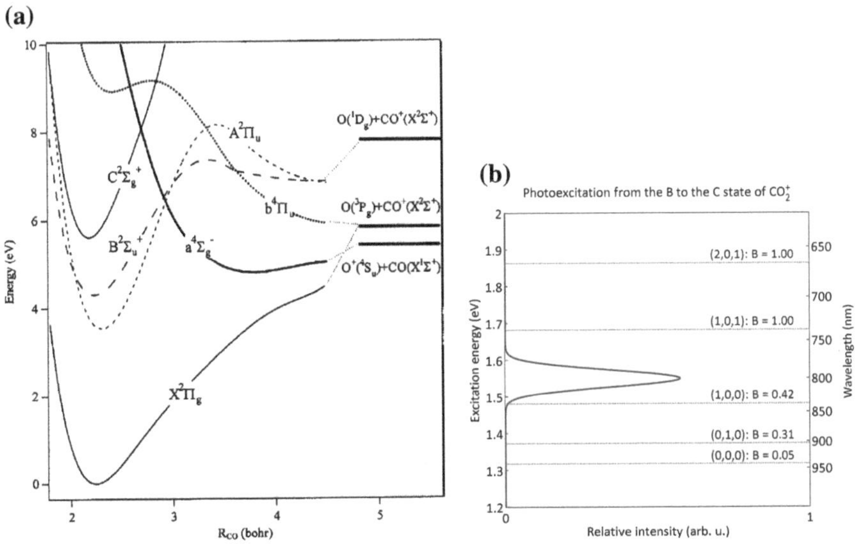

Fig. 7.1 **a** Potential energy curves involved in the predissociation mechanism of $CO_2^+(C^2\Sigma_g^+)$ as a function of the CO bond length. **b** Vibrational levels of the C state of CO_2^+ according to their excitation energies with respect to the B state in its ground vibrational state, denoted by the *horizontal dashed lines*. They are labled via their vibrational mode and branching ratio B for $CO^+_{(1,0)}$. In addition a gaussian spectrum centred at 800 nm with a FWHM of 40 nm is plotted. Data taken from [6]. (Part (a) reprinted with permission from [7]. Copyright (2003) by the American Institute of Physics.)

$$CO_2^+(b^4\Pi_u^+/^4B_1) \rightarrow CO_2^+(X^2\Pi_g) \rightarrow O(^3P_g) + CO^+(X^2\Sigma^+) \qquad (7.2)$$

where the first transition takes place via a conical intersection and is very fast. This channel is open in both linear and bent configuration of the molecule. However, when in bent mode a spin-orbit coupling between the $CO_2^+(^4B_1)$ and the $a^4\Sigma_g^-$ state becomes efficient. This opens two more pathways, as the $a^4\Sigma_g^-$ state may directly dissociate into $O^+(^4S_u)$ or couple to the ground state $CO_2^+(X^2\Pi_g)$:

$$CO_2^+(b^4\Pi_u^+/^4B_1) \rightarrow CO_2^+(a^4\Sigma_g^-) \rightarrow O^+(^4S_u) + CO(X^1\Sigma^+) \qquad (7.3)$$

$$CO_2^+(b^4\Pi_u^+/^4B_1) \rightarrow CO_2^+(a^4\Sigma_g^-) \rightarrow CO_2^+(X^2\Pi_g) \rightarrow O(^3P_g) + CO^+(X^2\Sigma^+) \qquad (7.4)$$

It has further been established that the corresponding fragments are released with near thermal kinetic energy distributions centred at zero kinetic energy release (KER) [14]. The reason for this is the non-Coulombic dissociation path and vibrational excitation of the diatomic fragment via the remaining excess energy.

7.1.2 Branching Ratio

The branching ratio R_b for one of the fragments, say $CO^+_{(1,0)}$, can be defined as follows:

$$R_b = \frac{CO^+_{(1,0)}}{CO^+_{(1,0)} + O^+_{(1,0)}} \tag{7.5}$$

In fixed wavelength (He I line) photoionisation studies, a branching ratio of $R_b \approx 0.2$ was found [10]. This agrees with theoretical calculations which propagate the relative vibrational state populations of the $CO_2^+(C^2\Sigma_g^+)$ through the predissociation mechanism (see Fig. 7.1b for the branching ratios of several vibrational states). It has been found however that $CO^+_{(1,0)}$ is favoured significantly when collision induced excitation takes place, resulting in a much higher branching ratio of about $R_b \approx 0.9$ [15]. The authors concluded that collision induced excitation would trigger dynamics exclusively in the linear configuration of CO_2^+, such that the dissociation pathway 7.3 is suppressed. This would require a significantly increased collision cross section for linear molecules. Here one may argue, however, that in a randomly aligned sample this effect would be unlikely to become dominant, so further reasons for the increased branching ratio should be considered. It might for example be possible that mainly the highest lying vibrational states ($\geq$19.76 eV) were populated through collision excitation, as these have been shown to exclusively produce $CO^+_{(1,0)}$ [6].

7.1.3 Excitation Pathways

Four pathways for reaching the third excited ionic state are possible and are illustrated in Fig. 7.2. Channel (1) denotes tunneling ionisation from the HOMO-3, which leaves the ion in the third excited state. Channels (2) and (3) consist of tunneling ionisation (TI) followed by photoexcitation. In (2) TI from HOMO-1 takes place and the resulting first excited ionic state is coupled to the third. The corresponding dipole transition would be perpendicular due to the symmetry of the involved molecular states. In (3) TI from HOMO-2 is followed by photoexcitation to the third excited state. Here, the dipole transition would be parallel. The last channel (4) consists of TI followed by an inelastic recollision. Here, TI from HOMO is displayed but lower lying orbitals may contribute as well.

In principle, all channels can contribute to the fragmentation yields. However, the three different excitation types can be distinguished by recording the fragmentation yields as a function of driving laser ellipticity and polarisation in the molecular frame. Here, recollision can be identified via its strong ellipticity dependence as demonstrated in the previous chapter. The remaining channels, TI and photoexcitation, may then be identified via their polarisation dependence. Here, (1) would display a higher fragmentation yield for parallel ionisation compared to perpendicular.

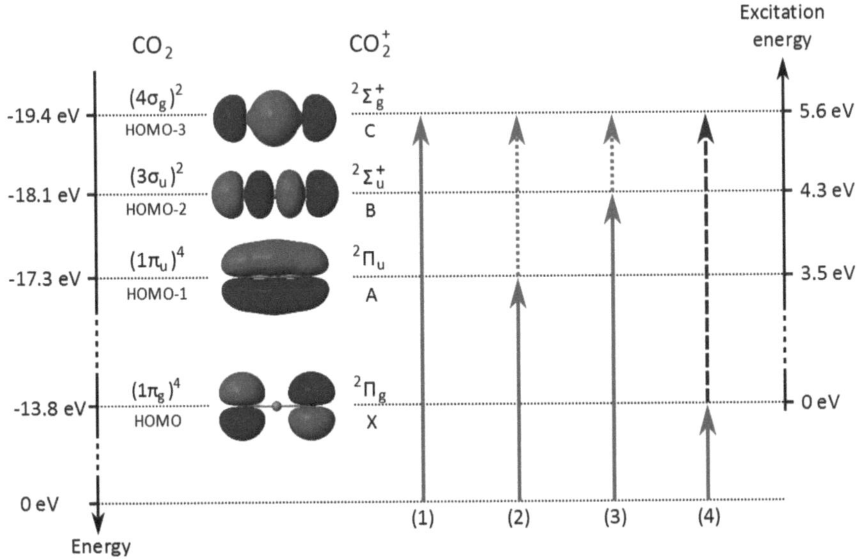

Fig. 7.2 Possible channels for reaching the third excited state of CO_2^+. *Solid arrows* denote tunneling ionisation, *dotted arrows* photo-excitation and *dashed arrows* electron recollision. (*1*) refers to tunneling ionisation from HOMO-3, leaving the ion in the third excited state. (*2*) illustrates ionisation from HOMO-1 followed by a perpendicular dipole transition. (*3*) illustrates tunneling ionisation from HOMO-2 followed by a parallel dipole transition and (*4*) depicts tunneling ionisation followed by recollision induced impact excitation

The difference can be expected to be on the order of 20 % as this was observed for TI from the HOMO of N_2 which has the same orbital symmetry [16]. Even though the TI step in channels (2) and (3) will introduce an angular dependence, one may expect that the polarisation dependence of the following dipole transitions dominates. In (2) the fragmentation yield would be proportional to $\sin^2(\theta)$ whilst (3) would vary as $\cos^2(\theta)$, where θ is the angle of the probe polarisation with respect to the molecular bond.

The DE channels have previously been investigated by McKenna et al. who have identified a strong enhancement of the fragmentation yield in linear, compared to circular polarisation [14]. This was attributed to recollision contributing significantly to the ion yields for linear polarisation. Here, one has to be cautious however, as all of the above described excitation channels have a significant polarisation dependence and hence may be strongly enhanced when compared to circular polarisation. The reason for this is that laser induced dynamic alignment of an unaligned sample in the rising edge of the laser pulse cannot be excluded even at low intensities in relatively long pulses (>10 fs) [17]. Furthermore a virtually aligned sample is always probed in the strong field regime, due to the scaling of the MO-ADK rate with the orientation of the molecule with respect to the linear laser polarisation. While the authors from [14] argue that alignment effects are negligible due to vanishing DE enhancement

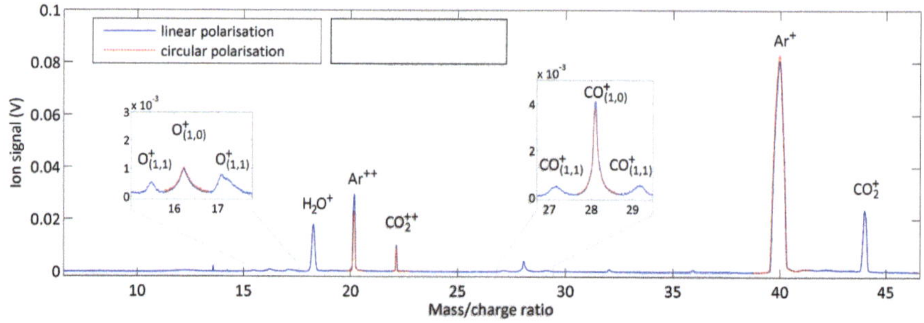

Fig. 7.3 TOF spectrum from a 10 % CO_2 in Ar mix at 2.1 ± 0.2 bar backing pressure. The driving laser pulse polarisation was perpendicular to the molecular axis at an average intensity of 5×10^{14} Wcm^{-2}. In order to provide a reference for the fragmentation yield variation as a function of linear polarisation angle, the yields for circular polarisation at matching Ar^+ rate recorded and superimposed on the spectrum (*red dotted line*)

at high intensities, this could also be attributed to barrier suppression rendering the molecule practically structureless and hence suppressing the polarisation dependence of excitation channels that rely on orbital symmetry.

7.2 Experiment

7.2.1 Setup and Procedure

The aim of the experiment was to identify and control the strong field induced dissociative excitation channels in CO_2^+. For this all previously discussed laser parameter dependences were studied in the molecular frame of reference, which was achieved by employing laser induced impulsive alignment. The resulting fragmentation yields were extracted from the TOF spectra obtained from the experiments presented in the previous chapter. A TOF spectrum is displayed in Fig. 7.3, where the O^+ and CO^+ peaks are magnified in the figure insets. Note that the peaks are split according to the kinetic energy release (KER) of the fragments, because the linear probe laser polarisation was parallel to the TOF axis. This implies that the central peak has a narrow KER distribution centred at zero KER. As was pointed out in the background section, this is a characteristic signature of the $O^+_{(1,0)}$ and $CO^+_{(1,0)}$ channels, which could thus be identified and isolated from other fragmentation channels. The setup, procedure and data analysis were discussed in detail in Sect. 6.3. In addition, the polarisation scans were extended by two reference measurements.

Firstly, the fragmentation yields were recorded using a circularly polarised probe for comparison with the yields from linear polarisation. However, it has been pointed out in Sect. 2.1.2 that the ionisation yield per pulse in a circularly polarised field is lower than in a linearly polarised one at the same intensity. For achieving the same ionisation rate in atoms and small molecules, the intensity relationship $I_{(lin)} = 0.65 I_{(cir)}$ has been verified [14, 18, 19]. Yet in this experimental setup, significant focal volume averaging is expected, such that one cannot rely on this exact relationship. In this experiment, the ionisation rate of Ar was measured in situ, such that the intensity for circular polarisation can be adjusted to match the Ar^+ rate in the linear case, thus establishing a meaningful comparison. This is displayed in Fig. 7.3. In practice, an intensity scan was conducted with the circularly polarised probe and the TOF spectrum selected where the Ar^+ yield from circular polarisation best matched the average Ar^+ yield from the polarisation scan with a linearly polarised probe. This reference measurement could not be performed with the 1,350 nm probe pulses. The reason for this was the limited maximum pulse energy in the interaction region that could be obtained from the setup employing 1,350 nm pulses (for details see Sect. 4.3.2). This meant that the intensity could not be increased high enough for circular polarisation to achieve the matching of the Ar^+ rate for linear polarisation.

Secondly, the polarisation scans were repeated with an elliptically polarised probe, where $\epsilon = 0.3$ was chosen. All other conditions were kept the same compared to the associated polarisation scan with a linearly polarised probe. This amount of ellipticity ensured that recollision effects were switched off (see Chap. 6), whilst the major axis of ellipticity still provided the dominant electric field direction. This enabled the investigation of the polarisation dependence under recollision free conditions. Note that a new reference yield at circular polarisation was measured for scans using an elliptical probe.

For all polarisation scans, the data was normalised by dividing the obtained fragmentation yields by the highest yield from each scan. When using 800 nm probes, the reference yield for circular polarisation was normalised by the same value and the yield difference $\Delta F = F_{lin} - F_{cir}$ calculated. Here, F_{lin} is the normalised fragmentation yield at linear (or elliptical) and F_{cir} the corresponding yield at circular polarisation.

7.2.2 Collection of Fragment Ions

Despite the narrow KER distribution of the $CO^+_{(1,0)}$ and $O^+_{(1,0)}$ fragment ions one needs to ensure that all ions are collected independently of the orientation of the molecular axis with respect of the TOF axis. For this, the fragmentation yield was recorded as a function of the alignment pulse polarisation with respect to the TOF axis. A 30 fs, 800 nm, circularly polarised probe pulse at an average intensity of 3×10^{14} Wcm^{-2} was used, such that the fragment ion ejection was only determined by the alignment laser polarisation. The resulting curves for CO_2^+, $CO^+_{(1,0)}$ and $O^+_{(1,0)}$ are displayed in Fig. 7.4. One observes that the CO_2^+ yield is mostly independent of θ, whilst the fragment yields are maximised at about $\theta = 45°$ and at most about 5 % lower at $\theta = 90°$.

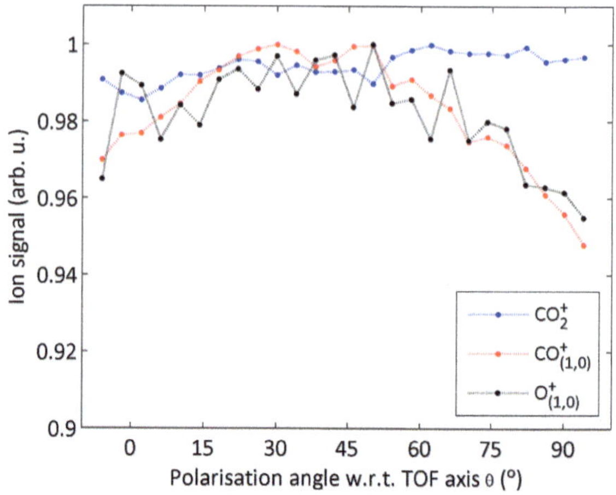

Fig. 7.4 CO_2^+, $CO^+_{(1,0)}$ and $O^+_{(1,0)}$ yield normalised to their respective maximum values as a function of polarisation angle θ. For the measurement, 800 nm, 30 fs, circularly polarised probe pulses at an average intensity of 3×10^{14} Wcm^{-2} were employed

The maximum variation observed for the fragment ions around their average yield over the probed angular range is approximately $\pm 2\,\%$. This uncertainty is inherent to the experimental setup. In the experiment, the observed variations must be larger than this uncertainty in order to be considered a significant change. It can be argued, however, that the observed uncertainty represents a special case and is expected to be lower for polarisation scans utilising linear or elliptical probe polarisation. For this one may note that the variation of the $CO^+_{(1,0)}$ rate does not display a random behaviour and is likely to be caused by a systematic error in the setup. It is unlikely that the maximum yield at $\theta = 45°$ is caused by an increased collection efficiency at this angle. This is because the collection efficiency for fragments with non-zero KER was found to decrease monotonically with increasing alignment angle θ with respect to the TOF axis (see Sect. 5.4.2). However, a maximum at this angular position can be caused by a non-perfect circular polarisation state. This is because the probe polarisation ellipse is rotated by the quarter-wave plate angle ϕ with respect to the incident linear probe polarisation. For a achieving circular polarisation, the major axis of the resulting polarisation ellipse thus takes an angle $\phi = 45°$ with respect to the TOF axis. If the polarisation ellipse is not perfectly circular and the recorded $CO^+_{(1,0)}$ yield has a strong polarisation dependence, this may lead to a systematic variation of the fragment signal around $\theta = 45°$. It is very likely that this effect is the cause for the observed variation in the fragmentation yield. In this case, this systematic error would be absent for linear probe polarisation.

7.3 Dissociative Excitation to $CO^+_{(1,0)}$

In this section the $CO^+_{(1,0)}$ channel is discussed. In particular, the intensity, ellipticity and polarisation dependence was studied for 800 and 1,350 nm probe pulses.

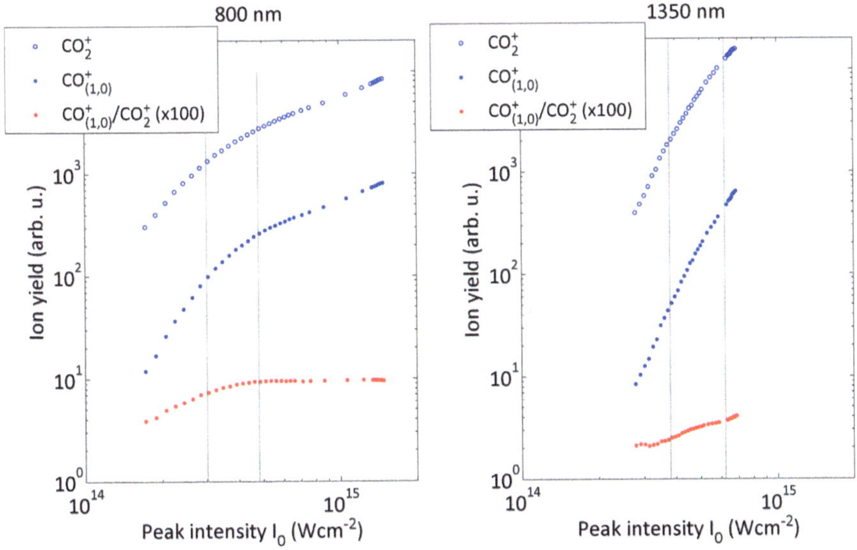

Fig. 7.5 CO_2^+, $CO^+_{(1,0)}$ and their ratio $CO^+_{(1,0)}/CO_2^+$ as a function of peak intensity I_0 for 30 fs probe pulses with a wavelength of 800 nm (*left panel*) and 1,350 nm (*right panel*). The *dotted vertical lines* denote the peak intensities employed for conducting ellipticity and polarisation measurements

7.3.1 Intensity Dependence

Results

Figure 7.5 displays the CO_2^+, $CO^+_{(1,0)}$ and their ratio $CO^+_{(1,0)}/CO_2^+$ as a function of peak intensity I_0 for 30 fs probe pulses with a wavelength of 800 and 1,350 nm. In both cases one observes that the single ionisation (SI) rate and the dissociation yield display a similar intensity dependence over the entire intensity regime. This manifests itself in a rather small positive slope of the $CO^+_{(1,0)}/CO_2^+$ ratio as a function of intensity. In the case of 800 nm the slope completely vanishes at peak intensities above 5×10^{14} Wcm^{-2}.

Discussion

The observed weak intensity dependence of the $CO^+_{(1,0)}/CO_2^+$ ratio is in accordance with the DE mechanism in CO_2^+. This is because CO_2^+ is always an intermediate state no matter which of the four channels leads to its third excited state. This implies that in this experiment, the intensity dependence of the $CO^+_{(1,0)}/CO_2^+$ ratio is not suitable for identifying the dominant excitation pathway for $CO^+_{(1,0)}$ production. Note however, that the vanishing slope of the ratio for high intensities at 800 nm suggests that both the SI and fragmentation channels are saturated. This is not observed for 1,350 nm in the probed intensity regime.

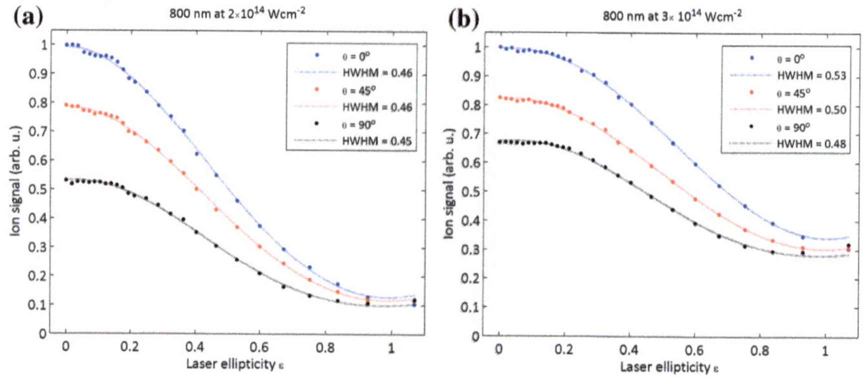

Fig. 7.6 Ellipticity scans for $CO^+_{(1,0)}$ at different values of θ using 30 fs, 800 nm probe pulses at average intensities of $I^{(lo)}_{800} = 2 \times 10^{14}$ Wcm^{-2} (**a**) and $I^{(hi)}_{800} = 3 \times 10^{14}$ Wcm^{-2} (**b**)

7.3.2 Ellipticity Dependence

Results

Figure 7.6 displays the ellipticity dependence of $CO^+_{(1,0)}$ formation in 30 fs, 800 nm pulses at average intensities of $I^{(lo)}_{800} = 2 \times 10^{14}$ Wcm^{-2} and $I^{(hi)}_{800} = 3 \times 10^{14}$ Wcm^{-2}. In each case, three angles of θ were probed, which denotes the angle between the major axis of the polarisation ellipse and the molecular axis. All curves are normalised with respect to the maximum fragmentation yield at $\epsilon = 0$ and $\theta = 0$. The half width at half maximum (HWHM) $\Delta\epsilon$ of each curve is obtained from the polynomial fits displayed in the figures. Here, very large values for $\Delta\epsilon^{(lo)}_{800} \approx 0.5$ and $\Delta\epsilon^{(hi)}_{800} \approx 0.5$ were extracted. In addition, one observes that a strong polarisation dependence is present, when comparing the relative yields for different values of θ at $\epsilon = 0$. One observes that $CO^+_{(1,0)}$ formation is most strongly pronounced for parallel polarisation and weakest for perpendicular polarisation. This effect is significantly stronger at $I^{(lo)}_{800}$ compared to $I^{(hi)}_{800}$. For both intensities, however, the polarisation dependence only vanishes when approaching circular polarisation at $\epsilon > 0.9$, where the curves for different values of θ converge to the same fragmentation yield. Here one should also note that the ion yield at circular polarisation compared to the maximum ion yield at $\epsilon = 0$ and $\theta = 0$, is about three times higher at $I^{(hi)}_{800}$ compared to $I^{(lo)}_{800}$.

Figure 7.7 displays the ellipticity dependence of $CO^+_{(1,0)}$ formation in 30 fs, 1,350 nm pulses at average intensities of $I^{(lo)}_{1350} = 2 \times 10^{14}$ Wcm^{-2} and $I^{(hi)}_{1350} = 4 \times 10^{14}$ Wcm^{-2}. The same qualitative features are observed for 1,350 nm compared to 800 nm probe laser radiation. Firstly similarly large values for the HWHM of the curves are extracted with $\Delta\epsilon^{(lo)}_{1350} \approx 0.5$ and $\Delta\epsilon^{(hi)}_{1350} \approx 0.6$. Yet for 1,350 nm, the fragmentation yield at circular polarisation relative to the largest yield at $\epsilon = 0$ and $\theta = 0$ is the same for both intensities. Furthermore, the same qualitative polarisation

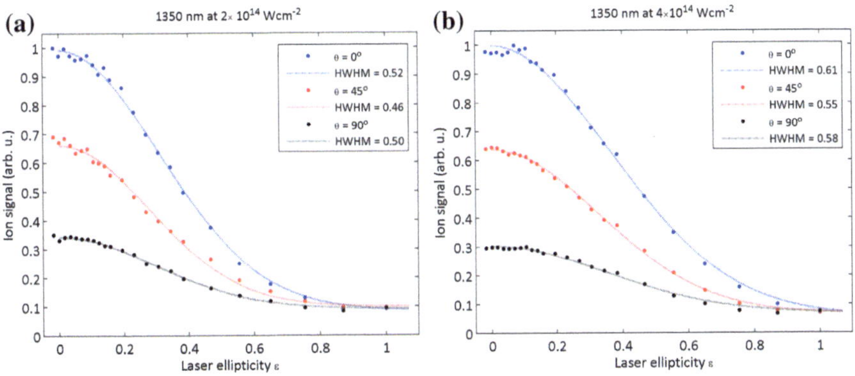

Fig. 7.7 Ellipticity scans for $CO^+_{(1,0)}$ at different values of θ using 30 fs, 1,350 nm probe pulses at average intensities of $I^{(lo)}_{1350} = 2 \times 10^{14}$ Wcm^{-2} (**a**) and $I^{(hi)}_{1350} = 4 \times 10^{14}$ Wcm^{-2} (**b**)

dependence is observed at $\epsilon = 0$ for the different values of θ. Compared to 800 nm, the observed polarisation dependence is more strongly pronounced for 1,350 nm radiation.

Discussion

First of all, the very large values extracted for $\Delta\epsilon$ show that inelastic recollisions do not play a role in the formation of $CO^+_{(1,0)}$ at any of the probed intensities or wavelengths. Furthermore one may even rule out a minor contribution of the associated excitation channel as all recorded ellipticity curves show a monotonous decrease over the entire ellipticity range and display a very weak ellipticity dependence for small ellipticities $\epsilon < 0.3$.

The fragment yield difference between parallel and perpendicular polarisation is larger than 30 % at all probed intensities and wavelengths. This suggests that the excitation mechanism is not dominated by TI from HOMO-3 (channel (1) in Fig. 7.2) as this would lead to a weaker polarisation dependence. In fact, the suppression of the $CO^+_{(1,0)}$ signal suggests that the main channel is TI from HOMO-2 followed by a parallel dipole transition. This is supported by the fact that the polarisation dependence only vanishes when the major axis of ellipticity becomes comparable to the minor axis, such that no dominant electric field direction is present anymore. This conjecture will be examined in more detail in the next section.

The observation that the polarisation dependence at 800 nm is reduced when the intensity is increased from $I^{(lo)}_{800}$ to $I^{(hi)}_{800}$ can be explained with the onset of ionisation saturation as observed in the intensity dependence of CO_2^+ and $CO^+_{(1,0)}$ at the same wavelength. In this case, over the barrier ionisation begins to contribute at the high intensity regions of the focus. In this ionisation process the electronic density profile of the molecular orbital does not play a role such that the polarisation dependence of the fragmentation yield vanishes. The onset of saturation also explains the increased relative yield at circular polarisation for higher intensities. As the signal is partially saturated, a decrease in average ionisation rate caused by changing the

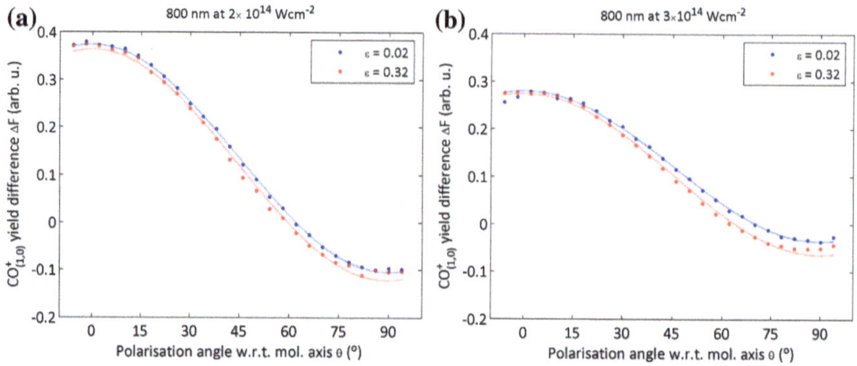

Fig. 7.8 Polarisation dependence for $CO^+_{(1,0)}$ at linear ($\epsilon = 0.02$) and elliptical ($\epsilon = 0.32$) polarisation using 30 fs, 800 nm probe pulses at average intensities of $I_{800}^{(lo)} = 2 \times 10^{14}$ Wcm^{-2} (**a**) and $I_{800}^{(hi)} = 3 \times 10^{14}$ Wcm^{-2} (**b**). Note that the plots display the difference ΔF between *linear* (or *elliptical*) and *circular* polarisation

pulse polarisation to circular has a reduced effect. These conclusions are consistent with the fact that the onset of ionisation saturation was not reached for 1,350 nm where the associated features of the ellipticity curves are not observed.

7.3.3 Polarisation Dependence

Results

Figure 7.8 displays the polarisation dependence for $CO^+_{(1,0)}$ at linear and elliptical polarisation using 30 fs, 800 nm probe pulses at average intensities of $I_{800}^{(lo)} = 2 \times 10^{14}$ Wcm^{-2} and $I_{800}^{(hi)} = 3 \times 10^{14}$ Wcm^{-2}. The yields are plotted as the difference ΔF between linear (or elliptical) and circular polarisation. For both intensities and employed values of ellipticity, the fragmentation yield decreases monotonically for increasing polarisation angle θ. Note that for both intensities, no significant difference is observed between the polarisation dependence at $\epsilon = 0.02$ and $\epsilon = 0.32$. Yet, the obtained curves differ in modulation depth and vertical offset with respect to zero yield difference. These two parameters were extracted from a numerical fit, where the function $F(\theta) = a \cos^2(c\theta - b) + d$ was fitted to the data. The goodness of the fit in terms of the R^2 parameter, the modulation depth a and the vertical offset d are displayed in Table 7.1 for each curve. Firstly, one observes that the polarisation dependence is very well described by the fit function due to high values of the associated R^2 parameter. Secondly, the modulation depth at $I_{800}^{(lo)}$ is significantly larger than at $I_{800}^{(hi)}$. Thirdly, the vertical offset is lower for $I_{800}^{(lo)}$ compared to $I_{800}^{(hi)}$. At both intensities, however, the modulation is not symmetric with respect to the reference yield from circular polarisation ($\Delta F = 0$).

Table 7.1 Characteristics of the different polarisation scans conducted for the $CO^+_{(1,0)}$ channel

λ_p (nm)	I_{av} (Wcm^{-2})	ϵ	ΔA	R^2	a	d
800	2×10^{14}	0.02	0.98	0.9997	0.48	−0.11
		0.32	0.97	0.9975	0.49	−0.12
	3×10^{14}	0.02	0.99	0.9995	0.32	−0.04
		0.32	1.01	0.9976	0.34	−0.06
1350	2×10^{14}	0.00	–	0.9984	0.67	0.31
	4×10^{14}	0.00	–	0.9973	0.68	0.29

ΔA denotes the ratio between the average Ar^+ yield at linear polarisation divided by the matched yield at circular polarisation. R^2 refers to the quality of the numerical fit to the data, with a and d being the modulation depth and vertical offset of the fitted function, respectively

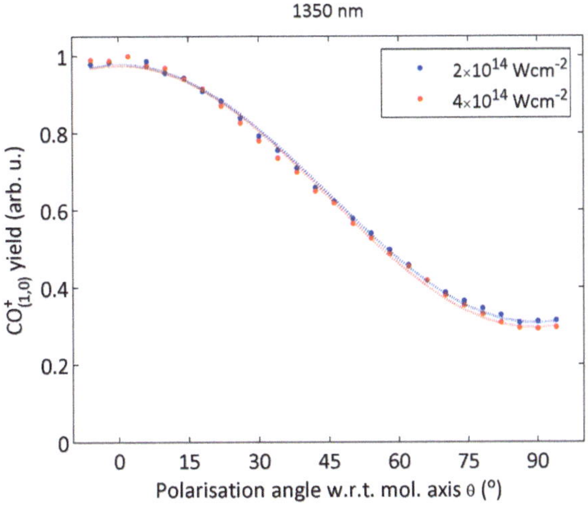

Fig. 7.9 Polarisation dependence for $CO^+_{(1,0)}$ at linear ($\epsilon = 0.00$) polarisation using 30 fs, 1,350 nm probe pulses at average intensities of $I^{(lo)}_{1350} = 2 \times 10^{14}$ Wcm^{-2} and $I^{(hi)}_{1350} = 4 \times 10^{14}$ Wcm^{-2}. Note that the curves display the fragmentation yields normalised with respect to the maximum value of each data set

Figure 7.9 displays the polarisation dependence for $CO^+_{(1,0)}$ at linear polarisation using 30 fs, 1,350 nm probe pulses at average intensities of $I^{(lo)}_{1350} = 2 \times 10^{14}$ Wcm^{-2} and $I^{(hi)}_{1350} = 4 \times 10^{14}$ Wcm^{-2}. Note that for each data curve, the normalised yield is plotted instead of the yield difference ΔF. Again, the function $F(\theta)$ was fitted to the data and the extracted parameters are displayed in Table 7.1. Firstly, one observes the same qualitative behaviour of the polarisation dependence for 1,350 nm compared to 800 nm. Due to the high values of the obtained R^2 parameter, it is very well described by the chosen fit function, which resulted in the same modulation depth and vertical offset for both intensities $I^{(lo)}_{1350}$ and $I^{(hi)}_{1350}$. Secondly, compared to the results from 800 nm, a significantly higher modulation depth can be observed at 1,350 nm.

Discussion

The major result of the polarisation measurement is that the mechanism for $CO^+_{(1,0)}$ production is dominated by a channel consisting of TI from HOMO-2 followed by a parallel dipole transition to the third excited ionic state in CO_2^+. This is because of the $\cos^2(\theta)$ dependence of the associated fragmentation yield. This finding is fully consistent with the interpretation of the ellipticity measurements. The polarisation scans show that this channel remains dominant over a large range of probe laser intensities and wavelengths. This can be explained by the very high intensities applied by the probe pulses. In this scenario, the potential energy surfaces of the ionic states are expected to be distorted and dressed, whilst the high photon numbers allow for a significant probability for Raman-type transitions of several orders to take place alongside single- and multiphoton transitions. This implies that the excitation bandwidth of the strong field coupling would be much broader than in a scenario where only single- or multiphoton transitions via 800 or 1,350 nm photons are available. One can observe, however, that the extracted modulation depth a is much larger in the case of 1,350 nm. With the above arguments it is unlikely that this can be attributed to a stronger coupling of the B and C state of CO_2^+ when using 1,350 nm radiation. Instead, this effect is most likely due to the significantly improved alignment quality present in the measurements employing the 1,350 nm probe pulse compared to the setup using 800 nm (see Table 6.2).

Furthermore it is observed that the polarisation dependences resulting from a linear and elliptical probe pulse are identical. This confirms the interpretation of the ellipticity measurements, such that recollision can be fully excluded from the DE mechanism in CO_2^+. This contradicts the results found by McKenna and coworkers [14] and possible reasons for this have been discussed in the background section. A possible reason for the absence of recollision contribution to the fragmentation mechanism can be found in the work of Rajgara and coworkers [20]. By investigating the contribution of recollision induced fragmentation in differently sized molecules, they concluded that this channel was most dominant for the fragmentation channels which require the highest energy transfer to the cation (see Fig. 2.19a). This is because the probability of laser induced multiphoton transitions (or tunneling ionisation) decreases nonlinearly with the excitation energy (or ionisation potential), whilst the probability for recollision excitation can be approximated to remain constant, as long as the maximum recollision energy is above the excitation energy. However, if the excitation energy is on the order of only a few driving laser photons, the high intensity in the strong field regime allows for a large probability of laser induced transitions. It is very likely that this is the reason for the $CO^+_{(1,0)}$ channel being dominated by a combination of tunneling ionisation and photoexcitation, whilst recollision plays negligible role.

The fact that the modulation depth of the polarisation dependence at 800 nm is reduced when increasing the intensity from $I_{800}^{(lo)} = 2 \times 10^{14}$ Wcm^{-2} to $I_{800}^{(hi)} = 3 \times 10^{14}$ Wcm^{-2} is most likely caused by ionisation saturation, analogously to the discussion of the ellipticity dependence in the previous section. This observation therefore does not contradict the previous interpretations. One should note, however, that the modulation of the fragmentation yield at 800 nm as a function of θ is not symmetric with respect to yield obtained from circularly polarised probe pulses ($\Delta F = 0$). This is an unexpected finding as the fragmentation yield from a circularly polarised field can be assumed to be an average over all polarisation directions and thus θ. Here one may add that the observed asymmetry is too large to be due to the circular polarisation state being slightly distorted as was observed in Sect. 7.2.2. The most likely explanation for this feature of the data may thus be that the matching of the Ar^+ rates is not an appropriate procedure for establishing the same ionisation rates for all ionisation channels. This is despite the good match achieved for each polarisation scan. For checking this, a matching factor was computed as $\Delta A = \langle A_{lin}\rangle / A_{cir}$, where $\langle A_{lin}\rangle$ denotes the average Ar^+ yield of the polarisation scan and A_{cir} the matched yield at circular polarisation. The resulting values for ΔA are displayed in Table 7.1. The reason for this could be the large focal volume averaging effect that has been avoided in experiments where ionisation rate matching for linear and circular polarisation has been achieved successfully [18, 19].

7.4 Dissociative Excitation to $O^+_{(1,0)}$

In this section the $O^+_{(1,0)}$ channel is discussed. In particular, the intensity, ellipticity and polarisation dependence was studied for 800 and 1,350 nm probe pulses and compared to the results for the $CO^+_{(1,0)}$ channel.

7.4.1 *Intensity Dependence*

Results

Figure 7.10 displays the CO_2^+, $O^+_{(1,0)}$ and their ratio $O^+_{(1,0)}/CO_2^+$ as a function of peak intensity I_0 for 30 fs probe pulses with a wavelength of 800 and 1,350 nm. In both cases one observes that the single ionisation (SI) rate and the dissociation yield display a similar intensity dependence over the entire intensity regime. This results in a small positive slope of the $O^+_{(1,0)}/CO_2^+$ ratio as a function of intensity. As the same was observed for the intensity dependence of $CO^+_{(1,0)}$ (see Fig. 7.5), this implies that the $CO^+_{(1,0)}$ and $O^+_{(1,0)}$ have approximately the same intensity dependence. Yet, the $O^+_{(1,0)}$ rate is about an order of magnitude lower than the $CO^+_{(1,0)}$ rate over the entire intensity range at both probe wavelengths.

Discussion

As in the case of $CO^+_{(1,0)}$, the observed weak intensity dependence of the $O^+_{(1,0)}/CO_2^+$ ratio shows that CO_2^+ is an intermediate state in the dissociation pathway of $O^+_{(1,0)}$. This implies that at intensities above 5×10^{14} Wcm^{-2} this dissociation channel is expected to begin to saturate, even though the slope of the $O^+_{(1,0)}/CO_2^+$ ratio is not observed to vanish completely in this intensity regime.

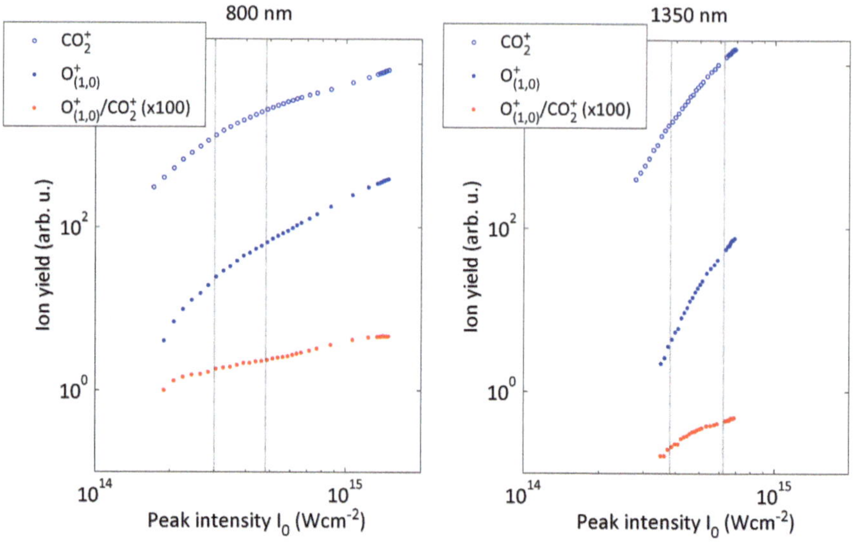

Fig. 7.10 CO_2^+, $O^+_{(1,0)}$ and their ratio $O^+_{(1,0)}/CO_2^+$ as a function of peak intensity I_0 for 30 fs probe pulses with a wavelength of 800 nm (*left panel*) and 1,350 nm (*right panel*). The *dotted vertical lines* denote the peak intensities employed for conducting ellipticity and polarisation measurements

7.4.2 Ellipticity Dependence

Results

Figure 7.11 displays the ellipticity dependence of $O^+_{(1,0)}$ formation in 30 fs, 800 nm pulses at average intensities of $I_{800}^{(lo)} = 2 \times 10^{14}$ Wcm^{-2} and $I_{800}^{(hi)} = 3 \times 10^{14}$ Wcm^{-2}. In each case, three angles were probed, $\theta = 0°, 45°, 90°$. All curves are normalised with respect to the maximum fragmentation yield at $\epsilon = 0$ and $\theta = 0°$. The half width at half maximum $\Delta\epsilon$ of each curve is obtained from the polynomial fits displayed in the figures. Similar to the results on $CO^+_{(1,0)}$, very large values for $\Delta\epsilon_{800}^{(lo)} \approx 0.4$ and $\Delta\epsilon_{800}^{(hi)} \approx 0.5$ were extracted. The polarisation dependence can be observed when comparing the $O^+_{(1,0)}$ rates at $\epsilon = 0$ for different values of θ. Here, the same qualitative—albeit less pronounced—behaviour as for $CO^+_{(1,0)}$ is displayed, with the rate decreasing with increasing angle θ. However, there is a new feature in the $O^+_{(1,0)}$ data: around $\epsilon = 0.4$, the ellipticity curves for $\theta = 45°$ and $\theta = 90°$ merge, whilst the curve at $\theta = 0°$ only converges to the remaining curves at ellipticities $\epsilon > 0.8$. This feature is observed for both intensities at 800 nm.

Figure 7.12 displays the ellipticity dependence of $O^+_{(1,0)}$ formation in 30 fs, 1,350 nm pulses at average intensities of $I_{1350}^{(lo)} = 2 \times 10^{14}$ Wcm^{-2} and $I_{1350}^{(hi)} = 4 \times 10^{14}$ Wcm^{-2}. At both intensities, similarly large values for the HWHM $\Delta\epsilon$ compared to 800 nm are extracted, with the exception of the curve associated

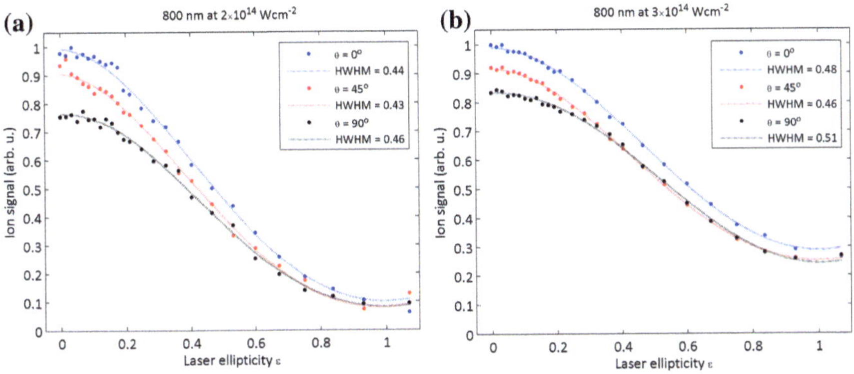

Fig. 7.11 Ellipticity scans for $O^+_{(1,0)}$ at different values of θ using 30 fs, 800 nm probe pulses at average intensities of $I^{(lo)}_{800} = 2 \times 10^{14}$ Wcm^{-2} (**a**) and $I^{(hi)}_{800} = 3 \times 10^{14}$ Wcm^{-2} (**b**)

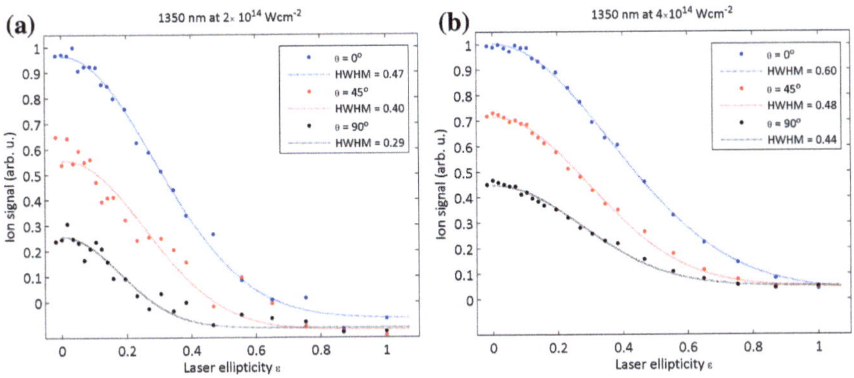

Fig. 7.12 Ellipticity scans for $O^+_{(1,0)}$ at different values of θ using 30 fs, 1,350 nm probe pulses at average intensities of $I^{(lo)}_{1350} = 2 \times 10^{14}$ Wcm^{-2} (**a**) and $I^{(hi)}_{1350} = 4 \times 10^{14}$ Wcm^{-2} (**b**).

with $\theta = 90°$ and $I^{(lo)}_{1350}$ where a significantly smaller value was obtained. The same qualitative polarisation dependence is displayed for 1,350 nm compared to 800 nm, yet it is much more pronounced for the longer wavelength. In contrast to the 800 nm case, the curves for different values of θ only merge for ellipticities $\epsilon > 0.8$.

Discussion

From the large values for the extracted HWHM $\Delta\epsilon$ one may infer that inelastic recollisions cannot be the dominant mechanism of $O^+_{(1,0)}$ formation at any of the probed wavelengths or intensities. In this context, the significantly lower value of $\Delta\epsilon \approx 0.29$ at $\theta = 90°$, 1,350 nm and $I^{(lo)}_{1350}$ represents a surprising exception. Whilst a physical effect cannot be excluded to cause an exceptionally strong ellipticity dependence at this alignment angle, it should also be mentioned that the corresponding data had the lowest count rate out of all conducted ellipticity measurements. This would imply

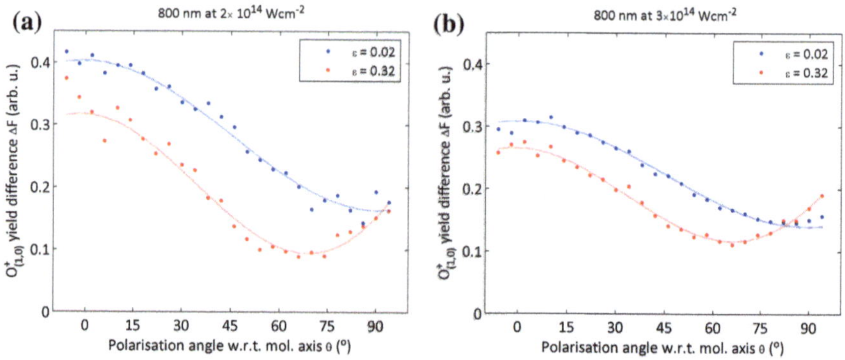

Fig. 7.13 Polarisation dependence for $O^+_{(1,0)}$ at linear ($\epsilon = 0.02$) and elliptical ($\epsilon = 0.32$) polarisation using 30 fs, 800 nm probe pulses at average intensities of $I^{(lo)}_{800} = 2 \times 10^{14}$ Wcm^{-2} (**a**) and $I^{(hi)}_{800} = 3 \times 10^{14}$ Wcm^{-2} (**b**). Note that the plots display the difference ΔF between *linear* and *circular* polarisation

that the uncertainty of the recorded data points is very large, such that the decrease of the $O^+_{(1,0)}$ rate with increasing ellipticity might not be quantified reliably through the employed polynomial fit.

A minor contribution from recollision excitation to the $O^+_{(1,0)}$ channel, however, cannot be ruled out. This is because at 800 nm the ellipticity curves for $\theta = 45°$ and $\theta = 90°$ merge at an ellipticity of approximately $\epsilon = 0.4$. Out of the four possible excitation mechanisms presented in Fig. 7.2, only the recollision excitation channel has a sufficiently strong ellipticity dependence to cause such a feature. Here, one has to note however that the feature is not observed for a 1,350 nm driving field. This is unexpected as the longer wavelength typically enhances recollision induced channels, as was demonstrated in the case for nonsequential double ionisation in the previous chapter. The possibility of a recollision contribution to the fragmentation yield will be investigated further when comparing polarisation scans for linear and elliptical probe pulses.

7.4.3 Polarisation Dependence

Results

Figure 7.13 displays the polarisation dependence for $O^+_{(1,0)}$ at linear and elliptical polarisation using 30 fs, 800 nm probe pulses at average intensities of $I^{(lo)}_{800} = 2 \times 10^{14}$ Wcm^{-2} and $I^{(hi)}_{800} = 3 \times 10^{14}$ Wcm^{-2}. The yields are plotted as the difference ΔF between linear and circular polarisation. For linear probe pulse polarisation one observes the same qualitative behaviour as for $CO^+_{(1,0)}$: the fragmentation yield decreases monotonically with increasing θ, is well described by a

Table 7.2 Characteristics of the different polarisation scans conducted for the $O^+_{(1,0)}$ channel

λ_p (nm)	I_{av} (Wcm^{-2})	ϵ	ΔA	R^2	a	d
800	2×10^{14}	0.02	0.98	0.9794	0.24	0.16
		0.32	0.97	–	–	–
	3×10^{14}	0.02	0.99	0.9941	0.17	0.14
		0.32	1.01	–	–	–
1350	2×10^{14}	0.00	–	0.9794	0.48	0.46
	4×10^{14}	0.00	–	0.9979	0.56	0.43

ΔA denotes the ratio between the average Ar^+ yield at linear polarisation divided by the matched yield at circular polarisation. R^2 refers to the quality of the numerical fit to the data, with a and d being the modulation depth and vertical offset of the fitted function, respectively

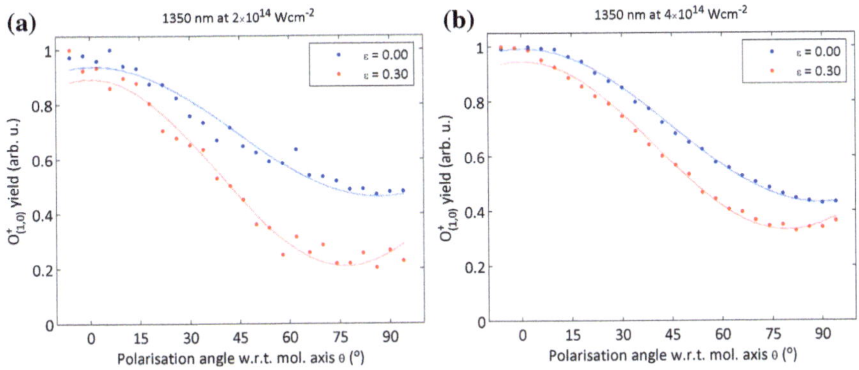

Fig. 7.14 Polarisation dependence for $O^+_{(1,0)}$ at linear ($\epsilon = 0.00$) and elliptical ($\epsilon = 0.30$) polarisation using 30 fs, 1,350 nm probe pulses at average intensities of $I^{(lo)}_{1350} = 2 \times 10^{14}$ Wcm^{-2} and $I^{(hi)}_{1350} = 4 \times 10^{14}$ Wcm^{-2}. Note that the curves display the fragmentation yields normalised with respect to the maximum value of each data set

fit function of the form $F(\theta) = a\cos^2(c\theta - b) + d$ (see Table 7.2 for the extracted fit parameters) and has a stronger modulation depth for $I^{(lo)}_{800}$ compared to $I^{(hi)}_{800}$. Quantitatively, however, there are two differences. Firstly, the modulation depth for $O^+_{(1,0)}$ production is half as large as the value for $CO^+_{(1,0)}$ for the same probe laser conditions. Secondly, the minimum value for the yield difference between linear and circular polarisation is significantly higher in the case of $O^+_{(1,0)}$ and always positive. For elliptical probe polarisation, a different polarisation dependence is observed at both intensities. The fragmentation yield decreases with increasing θ but reaches a local minimum at about $\theta_{min} = 65°$ and increases for $\theta > \theta_{min}$. Additionally, the yield difference ΔF is smaller for $\epsilon = 0.32$ compared to $\epsilon = 0.02$ for all angles $\theta < 75°$. As this behaviour cannot be described by the fit function $F(\theta)$, polynomials were fitted to the data in order to guide the readers eyes.

Figure 7.14 displays the polarisation dependence for $O^+_{(1,0)}$ at linear and elliptical polarisation using 30 fs, 1,350 nm probe pulses at average intensities of $I^{(lo)}_{1350} = 2 \times 10^{14}$ Wcm^{-2} and $I^{(hi)}_{1350} = 4 \times 10^{14}$ Wcm^{-2}. Note that for each data curve, the normalised yield is plotted instead of the yield difference ΔF. At linear probe polarisation, the same polarisation dependence is observed as in the 800 nm case and is well described by a $\cos^2(\theta)$ behaviour. The extracted modulation depth for $I_{1350}^{(lo)}$ is slightly lower than for $I_{1350}^{(hi)}$, yet both are about twice as large compared to the values obtained in the corresponding 800 nm cases. For elliptically polarised probe pulses, the polarisation dependence does not fulfill a $\cos^2(\theta)$ behaviour. Instead, a local minimum is formed at about 75°. Furthermore, the normalised $O^+_{(1,0)}$ yield decreases faster with increasing angle θ at elliptical compared to linear polarisation.

Discussion

First of all one may conclude that the $O^+_{(1,0)}$ channel is dominated by an excitation pathway consisting of tunneling ionisation from HOMO-2 and a parallel dipole transition to the third excited ionic state. This is supported by the $\cos^2(\theta)$ behaviour that describes the polarisation dependence of the $O^+_{(1,0)}$ channel very well at all intensities and wavelengths for linear probe polarisation. This is consistent with the results for the $CO^+_{(1,0)}$ channel, where the same dominant excitation pathway was found. This is because both fragmentation outcomes are expected to result from the same $CO_2^+(C^2\Sigma_g^+)$ state. One should note, however, that the modulation depth of the polarisation dependence of the $O^+_{(1,0)}$ channel is significantly weaker than that of the $CO^+_{(1,0)}$ channel at the same driving laser parameters. This suggests that the coupling strength of the dipole transition is weaker in the mechanism for $O^+_{(1,0)}$ production. From the predissociation mechanism of $CO_2^+(C^2\Sigma_g^+)$ it is known that $O^+_{(1,0)}$ production proceeds either via dissociation of the vibrational ground state (see pathway 7.1) or via pathway 7.3 which requires a bent configuration of the molecular ion. The vibrational population of the final state during an electronic transition is typically approximated to be independent of the electronic degrees of freedom and calculated via the associated Franck-Condon principle. One can thus assume that the coupling strength of the dipole transition between the B and C state of CO_2^+ is the same for all associated vibronic transitions. This implies that pathway 7.1 does not explain the reduced modulation depth. For a bent molecule, however, the orbital symmetry is altered which implies a modification of the transition rules for the allowed dipole transitions. It is thus very likely that the reduced modulation depth for the polarisation dependence of $O^+_{(1,0)}$ is an indicator for a large contribution of pathway 7.3 to the total $O^+_{(1,0)}$ yield in this experiment. It needs to be pointed out, however, that the above arguments neglect the influence of the strong field on the predissociation mechanism. In order to unambiguously determine the reason for the observed reduced coupling strength of the dipole transition for $O^+_{(1,0)}$ production, one would have to calculate the molecular dynamics within the field-dressed potential energy surfaces of CO_2^+ and include the vibrational degrees of freedom. In this context, one should also note that the reduced modulation depth of the $O^+_{(1,0)}$ yield when comparing the results for $I_{800}^{(lo)}$ and $I_{800}^{(hi)}$ is most likely due to ionisation saturation at the higher intensity as was concluded in the case of $CO^+_{(1,0)}$. The increased modulation depth when comparing

800–1,350 nm, however, is most likely due to the improved alignment quality when using 1,350 nm probe pulses. This is supported by the same observation in $CO^+_{(1,0)}$.

The observations made for the polarisation dependence resulting from elliptical probe pulses strongly suggest that the mechanism for $O^+_{(1,0)}$ production differs significantly from the mechanism for $CO^+_{(1,0)}$ formation. However, the reason for this cannot be determined unambiguously on the basis of the obtained results. First of all it needs to be stressed, that the effect of the ellipticity on the polarisation dependence of $O^+_{(1,0)}$ is unlikely to be due to an artefact in the measurements. This is because for 800 nm, where the effect is pronounced most strongly, the same change of the polarisation dependence is observed in the ellipticity scans. This shows that the effect can be reproduced via independent measurements. Additionally the effect is not observed for $CO^+_{(1,0)}$, which rules out a distorted polarisation state as a cause.

One may now argue that the change of the polarisation dependence is most likely due to contributions from recollision excitation of the parent ion (channel (4) in Fig. 7.2). This is because it is the only excitation pathway with an ellipticity dependence that is sufficiently strong to cause a significant change when increasing the ellipticity from $\epsilon = 0$ to 0.3. However, recollision excitation is expected to become less pronounced for higher intensities, as laser induced channels are enhanced non-linearly. This is not observed when comparing the polarisation dependence for elliptical probe pulses at $I^{(lo)}_{800}$ and $I^{(hi)}_{800}$. Furthermore, recollision induced channels are expected to become more pronounced at longer driving wavelengths. Yet the change in polarisation dependence for elliptical probe pulses is observed to be less pronounced at 1,350 nm compared to 800 nm. One thus has to conclude that recollision excitation is unlikely to be the cause of the observed effect, because this explanation is not consistent with all of the available experimental evidence.

Therefore, one has to consider that the reason for the change in polarisation dependence may be rooted in the predissociation pathway of $CO_2^+(C^2\Sigma_g^+)$ leading to $O^+_{(1,0)}$ fragments. Here, it was pointed out earlier that in this experiment $O^+_{(1,0)}$ production most probably requires the molecular ion to be in a bent configuration. The broken cylindrical symmetry of the resulting molecular orbital may then be related to the observed ellipticity effect. Nevertheless this is a speculation and for investigating this further, numerical modelling of the predissociation mechanism in a strong field would be required. However, such a theoretical treatment is beyond the scope of the current work.

7.5 Branching Ratio

In this section the branching ratio R_b for the $CO^+_{(1,0)}$ channel is investigated in order to better characterise the predissociation mechanism under the employed experimental conditions.

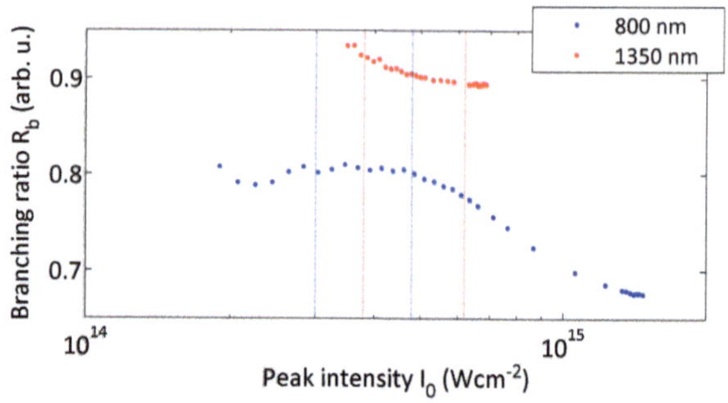

Fig. 7.15 Branching ratio R_b for $CO^+_{(1,0)}$ as a function of peak intensity I_0 for 30 fs probe pulses with a wavelength of 800 and 1,350 nm. The *dotted vertical lines* denote the peak intensities employed for conducting ellipticity and polarisation measurements

7.5.1 Intensity Dependence

Results

Figure 7.15 displays the intensity dependence of the branching ratio for $CO^+_{(1,0)}$ in 30 fs laser pulses at 800 and 1,350 nm. For 800 nm an approximately constant value of $R_b = 0.8$ is obtained for peak intensities $I_0 < 5 \times 10^{14}$ Wcm^{-2}. For higher intensities, the branching ratio monotonically decreases to about $R_b = 0.7$ at 1×10^{15} Wcm^{-2}. At 1,350 nm the branching ratio is approximately constant over the probed intensity range with a value of $R_b = 0.9$. For low intensities $I_0 < 5 \times 10^{14}$ Wcm^{-2} the branching ratio decreases slightly with increasing intensity.

Discussion

The main result of this measurement is the very high value of the branching ratio obtained at the intensities that were employed for the ellipticity and polarisation measurements. This implies that $CO^+_{(1,0)}$ production strongly dominates the predissociation mechanism in the strong field regime, whilst this feature is mostly independent of driving laser intensity and wavelength. The obtained branching ratio is very similar to particle collision-excitation experiments [21] but in stark contrast with fixed wavelength (He I line) photoionisation studies where a branching ratio of $R_b \approx 0.2$ was found.

There are two factors that can cause the $CO^+_{(1,0)}$ channel to dominate over the $O^+_{(1,0)}$ channel. Firstly, the main excitation pathway consisting of tunneling ionisation followed by a parallel dipole transition may preferentially lead to vibrational excitation in the $CO_2^+(C^2\Sigma_g^+)$ state such that pathway 7.1 is suppressed. Secondly, $O^+_{(1,0)}$ formation may be suppressed due to the molecular ion mostly taking a linear configuration. In this case dissociation pathway 7.3 would not be accessible. It is likely that both factors play a role in this experiment. In Sect. 7.4.3 it was argued that channel 7.3 is most likely the dominant pathway for $O^+_{(1,0)}$ production in this experiment. This implies that pathway 7.1 is indeed suppressed with the dipole transition to $CO_2^+(C^2\Sigma_g^+)$ preferentially causing vibrational excitation. With only pathway 7.3 remaining and requiring a bent configuration of the molecular ion, one may conclude that CO_2^+ is mostly formed in the linear configuration with only a small amount of ions being present in the bent configuration. The fact that the branching ratio is

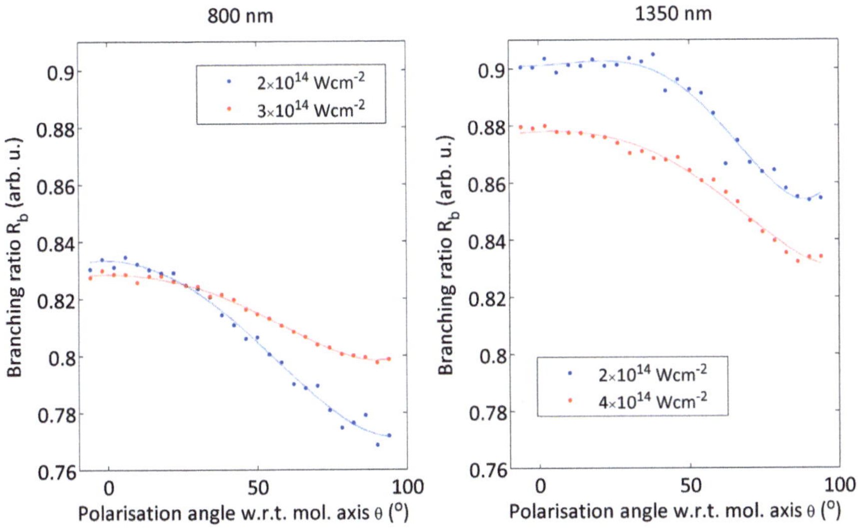

Fig. 7.16 Branching ratio R_b for $CO^+_{(1,0)}$ as a function of polarisation angle θ for 30 fs linearly polarised probe pulses with a wavelength of 800 nm (*left panel*) and 1,350 nm (*right panel*)

slightly higher for 1,350 nm may then be due to one or both of the above factors being pronounced more strongly for a longer driving laser wavelength.

The decreasing branching ratio for peak intensities higher than 5×10^{14} Wcm^{-2} at 800 nm is most probably caused by ionisation saturation, as this was observed to effect the fragmentation yield in this intensity regime.

7.5.2 Polarisation Dependence

Results

Figure 7.16 displays the polarisation dependence of the branching ratio R_b for $CO^+_{(1,0)}$ production for linearly polarised, 30 fs laser pulses at 800 and 1,350 nm. For 800 nm a branching ratio between 0.83 and 0.77 is observed that is monotonically decreasing for increasing polarisation angle θ. In the case of $I^{(lo)}_{800} = 2 \times 10^{14}$ Wcm^{-2}, a decrease of 0.06 is observed over the angular range, while this effect is only half as large at $I^{(hi)}_{800} = 3 \times 10^{14}$ Wcm^{-2}.

For 1,350 nm, the same qualitative behaviour is observed with the branching ratio taking slightly higher values compared to the 800 nm case between 0.9 and 0.83. The decrease over the angular range is about 0.4 at both employed intensities $I^{(lo)}_{1350} = 2 \times 10^{14}$ Wcm^{-2} and $I^{(hi)}_{1350} = 4 \times 10^{14}$ Wcm^{-2}.

Discussion

First of all one should note that the observed ranges for the branching ratio R_b at different wavelengths and intensities are consistent with the results from the intensity scans. Secondly, the observed decrease of the branching ratio for increasing θ is very small with a maximum decrease of about 6 % at 800 nm and $I_{800}^{(lo)}$. This is because the $CO^+_{(1,0)}$ and $O^+_{(1,0)}$ channels have the same qualitative polarisation dependence. The small decrease is then due to the stronger suppression of $CO^+_{(1,0)}$ at $\theta = 90°$ which was interpreted as being caused by a stronger coupling in the associated parallel dipole transition. Here, the smaller variation of the branching ratio at $I_{800}^{(hi)}$ must be attributed to the onset of ionisation saturation.

7.6 Conclusion

In the experiments presented in this chapter the dominating mechanism for dissociative excitation of the CO_2^+ ion in the strong field regime was identified and controlled. It was found that the corresponding pathway consists of two steps. Firstly, tunneling ionisation from the HOMO-2 of CO_2 takes place, leaving the parent ion in the second excited state B. Secondly, this state is coupled to the third excited ionic state C via a parallel dipole transition. This state then fully predissociates into $CO^+_{(1,0)}$ or $O^+_{(1,0)}$. When probed in the molecular frame of reference the parallel transition could be controlled via the polarisation, such that fragmentation yields were suppressed by up to 70 %. Furthermore, inelastic recollisions could be ruled out as part of the dissociation mechanism, due to the weak ellipticity dependence observed for the fragment ions. It was suggested that this is due to the low excitation energy required for reaching the relevant dissociation limits, such that laser induced channels dominate at the high intensities employed in the experiment.

It was then found that the dissociation mechanism preferentially produces $CO^+_{(1,0)}$ ions with a very large branching ratio between $R_b = 0.8$ and 0.9. It was concluded that this required the molecular parent ion to be mostly prepared in vibrationally excited states and to remain in a linear configuration during the predissociation mechanism. It is the above insights that render the observed molecular dynamics especially interesting from a theoretical point of view. Firstly, the excitation pathway offers the possibility to study the ultrafast population transfer between electronically coupled states in the strong field regime. Secondly, the predissociation mechanism is sensitive to the vibrational states and geometrical configuration of the dissociating molecular states. For this, theoretical efforts lead by Olga Smirnova and Misha Ivanov are currently underway modelling the obtained experimental results. This is expected to lead to a more quantitative understanding of molecular dynamics in the strong field regime and to test the interpretations of the dissociation mechanism presented in this chapter.

References

1. A.H. Zewail, Laser femtochemistry. Science **242**, 1645–1653 (1988)
2. H.J. Wörner, J.B. Bertrand, B. Fabre, J. Higuet, H. Ruf, A. Dubrouil, S. Patchkovskii, M. Spanner, Y. Mairesse, V. Blanchet, E. Mvel, E. Constant, P. B. Corkum, and D. M. Villeneuve, Conical intersection dynamics in No_2 probed by homodyne high-harmonic spectroscopy. Science **334**, 208–212 (2011)
3. P. Hockett, C.Z. Bisgaard, O.J. Clarkin, A. Stolow, Time-resolved imaging of purely valence-electron dynamics during a chemical reaction. Nat. Phys. **7**, 612–615 (2011)
4. A.E. Boguslavskiy, J. Mikosch, A. Gijsbertsen, M. Spanner, S. Patchkovskii, N. Gador, M.J.J. Vrakking, A. Stolow, The multielectron ionization dynamics underlying attosecond strong-field spectroscopies. Science **335**, 1336–1340 (2012)
5. F. Krausz, M. Ivanov, Attosecond physics. Rev. Mod. Phys. **81**, 163 (2009)
6. R. Locht, The dissociative ionization of carbon dioxide below 21.22ev. a high resolution pipeco spectroscopic investigation. Int. J. Mass Spectrom. Ion Processes **148**(1–2), L17–L23 (1995)
7. J. Liu, W. Chen, M. Hochlaf, X. Qian, C. Chang, C.Y. Ng, Unimolecular decay pathways of state-selected Co_2^+ in the internal energy range of 5.2–6.2 eV: an experimental and theoretical study. J. Chem. Phys. **118**, 149–163 (2003)
8. J. Eland, Predissociation of triatomic ions studied by photo-electron-photoion coincidence spectroscopy and photoion kinetic energy analysis: I. Co_2^+. Int. J. Mass Spectrom. Ion Phys. **9**(4), 397–406 (1972)
9. M.T. Praet, J.C. Lorquet, G. Raseev, Unimolecular reaction paths of electronically excited species. iv. the $^2\sigma_g^+$ state of Co_2^+. J. Chem. Phys. **77**(9), 4611–4618 (1982)
10. R. Bombach, J. Dannacher, J.-P. Stadelmann, J.C. Lorquet, Branching ratios and partition of the excess energy for the predissociation of CO_2^+ $\tilde{C}^2\Sigma_g^+$ molecular cations. J. Chem. Phys. **79**, 4214 (1983)
11. R. Locht, M. Davister, The dissociative electroionization of carbon dioxide by low-energy electron impact. the C^+, O^+ and Co^+ dissociation channels. Int. J. Mass Spectrom. Ion Processes **144**(1–2), 105–129 (1995)
12. Q. Meng, M.-B. Huang, H.-B. Chang, Theoretical study on the predissociation mechanism of Co_2^+. J. Phys. Chem. A **113**, 12825–12830 (2009)
13. M. Yang, L. Zhang, L. Lai, D. Zhou, J. Wang, Q. Sun, Study on the [1 + 1'] photodissociation spectra of Co_2^+ via transitions. Chem. Phys. Lett. **480**, 41–45 (2009)
14. J. McKenna, M. Suresh, B. Srigengan, I.D. Williams, W.A. Bryan, E.M.L. English, S.L. Stebbings, W.R. Newell, I.C.E. Turcu, J.M. Smith, E.J. Divall, C.J. Hooker, A.J. Langley, J.L. Collier, Rescattering-enhanced dissociation of a molecular ion. Phys. Rev. A **74**, 043409 (2006)
15. V.R. Bhardwaj, D. Mathur, K. Vijayalakshmi, P. Hvelplund, M.O. Larsson, C.P. Safvan, Collision-induced dissociation of Co_2^+ ions. Phys. Rev. A **58**(4), 2834–2843 (1998)
16. I. Litvinyuk, Alignment-Dependent strong field ionization of molecules. Phys. Rev. Lett. **90**, 233003 (2003)
17. A.S. Alnaser, C.M. Maharjan, X.M. Tong, B. Ulrich, P. Ranitovic, B. Shan, Z. Chang, C.D. Lin, C.L. Cocke, I.V. Litvinyuk, Effects of orbital symmetries in dissociative ionization of molecules by few-cycle laser pulses. Phys. Rev. A **71**, 031403 (2005)
18. M. Suresh, J. McKenna, B. Srigengan, I. Williams, E. English, S. Stebbings, W. Bryan, W. Newell, E. Divall, C. Hooker, A. Langley, Multiple ionization of ions and atoms by intense ultrafast laser pulses. Nucl. Instrum. Methods Phys. Res. Sect. B Beam Interact. Mater. Atoms **235**(1–4), 216–220 (2005)
19. J. McKenna, M. Suresh, B. Srigengan, I.D. Williams, W.A. Bryan, E.M.L. English, S.L. Stebbings, W.R. Newell, I.C.E. Turcu, J.M. Smith, E.J. Divall, C.J. Hooker, A.J. Langley, J.L. Collier, Ultrafast ionization study of N_2 in intense linearly and circularly polarized laser fields. Phys. Rev. A **73**(4), 043401 (2006)

20. F.A. Rajgara, M. Krishnamurthy, D. Mathur, Electron rescattering and the dissociative ionization of alcohols in intense laser light. J. Chem. Phys. **119**(23), 12224 (2003)
21. D. Mathur, A.K. Dharmadhikari, F.A. Rajgara, J.A. Dharmadhikari, Molecular symmetry effects in the ionization of Cs_2 by intense, few-cycle laser pulses, 0801.2046 (2008)

Chapter 8
Conclusion

In this chapter, the results obtained from the experiments presented in this thesis are summarised and discussed in a broader context. Their contribution to the scientific community is pointed out along with possible improvements and an outlook concerning possible future research.

8.1 Chapter 5: Characterisation and Optimisation of Molecular Alignment

In the experiments presented in this chapter, the revival structure of rotational wavepackets induced in the molecules N_2, CO_2 and O_2 was recorded via Coulomb explosion combined with a TOF detection technique. In such a scheme, the data only represents a qualitiative measure of the associated alignment distribution of the molecular ensemble. Because of this, a procedure was developed which accurately matched numerical simulations to the experimental data for extracting the alignment distribution in form of the expectation value $\langle\cos^2\theta\rangle$ and the rotational temperature of the molecular ensemble. This was done in the Fourier domain of the temporal evolution of the rotational wavepacket. The procedure was then applied to optimise the quality of molecular alignment within the employed setup. Here, the molecular sample was seeded in Ar as a carrier gas and the mixing ratio adjusted for obtaining the lowest possible rotational temperature, which was limited by the onset of clustering in the gas jet. The enhanced cooling lead to an improved degree of molecular alignment, which was optimised for a molecular sample concentration of 10 %. Compared to the pure sample N_2 was cooled from 24 ± 4 to 9 ± 2 K corresponding to an increase of $\langle\cos^2\theta\rangle$ from 0.60 to 0.71. CO_2 was cooled from 34 ± 3 to 9 ± 2 K improving $\langle\cos^2\theta\rangle$ from 0.48 to 0.64. For O_2 an enhanced cooling from 58 ± 2 to 37 ± 4 K was observed with an increase of $\langle\cos^2\theta\rangle$ from 0.49 to 0.58.

The relevance of the obtained results in the context of molecular physics is twofold. Firstly, the developed procedure for the characterisation of the rotational wavepacket

M. Oppermann, *Resolving Strong Field Dynamics in Cation States of CO_2 via Optimised Molecular Alignment*, Springer Theses, DOI: 10.1007/978-3-319-05338-7_8,

evolution represents the first systematic method that can be applied for any probing scheme that results in a linearly scaled measure of $\langle\cos^2\theta\rangle$. This is the case for TOF spectrometry as in this thesis and high harmonic generation, for example. It thus allows for the characterisation of molecular alignment in situ for such experiments without the need for additional setups. Secondly, the results on optimising the alignment quality via mixed gases suggest a route for improving the alignment quality in experimental setups that are limited in backing pressure. Whilst this approach is not novel, a systematic optimisation concerning the mixing ratio has not been reported previously. Hence the obtained results provide a roadmap for achieving very high degrees of molecular alignment with rather simple means.

Future developments motiviated by this research could go in two directions. Firstly, it is always desirable to increase the degree of molecular alignment even further. When mixed in Ar, this is not possible within this setup due to the limit imposed by clustering. It would therefore be interesting to repeat the experiments with Ne as a carrier gas, as it is known to cluster at higher backing pressures. In this context, it would furthermore be desirable to increase the applicable backing pressure of the molecular beam machine. However, this would require a different gas valve architecture and improved pumping speed of the vacuum system. The latter has been achieved recently by installing a second rotary vane pump for the backing of the source chamber turbo-molecular pumps. Secondly, it is well known that some molecules display higher order revivals [1]. This has also been observed in the experiments presented in this thesis in the case of CO_2 (the corresponding data was not displayed or discussed). Within the employed detection scheme, the revival structure can be observed in different fragmentation channels. It was found that the higher order revivals display different behaviour for different fragmentation channels. Yet, for further conclusions, this effect needs to be studied more systematically. The corresponding experiments have been performed recently and the analysis is currently underway.

8.2 Chapter 6: Recollision Induced Double Ionisation

In this chapter the molecular structure dependence of recollision induced double ionisation was studied. Here, it is well known that the orbital symmetry of the contributing molecular states is encoded into the angular dependent signature of the process. This signature was resolved by employing impulsive molecular alignment and enabled the identification of the specific multichannel contributions to the underlying mechanism for the first time. Using CO_2 as an example, the wavelength of the strong driving laser field was increased from 800 to 1350 nm at $2 \times 10^{14}\,\mathrm{Wcm}^{-2}$. This established nonsequential double ionisation (NSDI) as the exclusive mechanism for CO_2^{++} formation. Under these conditions the recollision event was completely characterised by measuring the shape of the recolliding electron wavepacket and the inelastic electron-ion recollision cross section. The obtained results unambiguously reveal the contribution from both the ionic ground and first excited state to the

NSDI mechanism in CO_2. Additionally, the experiment represents the first angularly resolved electron-ion collision measurement in CO_2, albeit over a finite collision energy range.

First of all, the results obtained in this chapter are relevant to the scientific community due to the novel insights into the molecular dynamics in the NSDI mechanism. It is well known that lower lying orbitals contribute significantly to single ionisation yields and play an important role in strong field phenomena such as high harmonic generation. The population dynamics of the associated ionisation channels are of special interest as they are at the heart of phenomena such as charge migration and molecular dissociation in the strong field regime. Studying the molecular dynamics inherent in the NSDI mechanism thus represented a gap that needed to be addressed, especially because the contribution of specific ionic channels in molecular NSDI had not been identified previously. The advantage of the employed experimental scheme for extracting this information is that it is general and can be applied to a range of small linear molecules. Furthermore, with TOF spectrometry it relies on the simplest possible ion detection scheme. This enables fast data acquisition rates and a robust setup. The disadvantage is that the scheme does not allow for the retrieval of specific branching ratios for the identified NSDI channels, as the electronic states of CO_2^{++} cannot be resolved. Furthermore it was not possible to distinguish the two possible mechanisms for recollision induced ionisation: direct impact ionisation and recollision excitation with subsequent field ionisation (RESI). This has only been achieved via COLTRIMS setups. To the strong field community, this research is also relevant as it establishes a roadmap for using inelastic electron recollisions for probing and inducing molecular dynamics. For this, a long driving laser wavelength is required in order to enhance recollision induced channels at the expense of direct laser induced ones. In this scenario, the induced molecular dynamics can be controlled by manipulating the recollision event. In the present experiment, this was achieved by varying the recollision angle with respect to the molecular axis.

Future developments informed by the obtained results may go several directions. First of all, one particular shortcoming of the employed setup was the limited pumping speed of the vacuum system. This imposed an upper limit of 47 Hz on the gas jet repetition rate and thus limited the data acquisition rate. However, through the use of an additional backing pump for the source chamber turbos, the pumping speed has been improved recently enabling a gas jet repetition rate of 300 Hz. Secondly, the employed experimental scheme should be applied to other molecules such as N_2 and O_2. Within this setup, another interesting experiment is possible. Through the use of elliptical probe pulses ($\epsilon = 0.3$) recollision may be switched off, while maintaining a dominant polarisation direction. This would enable the study of the angular signature of sequential double ionisation (SDI), which has not been reported thus far. It would be particularly interesting to compare the involvement of intermediate ionic states in the NSDI and SDI mechanisms. A third direction of research motivated by the presented results would be related to modifications of the experimental scheme. For instance, the recollision time was not controlled in this experiment. This can be achieved by the use of perpendicular two-colour driving laser fields, that effectively allow for the selection of specific recollision trajectories through the phase shift between the fields. This would in principle allow for subcycle time-resolution of NSDI. The selection of specific electron trajectories would also imply more precise control over the recollision energy, without changing the laser parameters. This could be a novel way to distinguish between direct impact and RESI channels for example. This is because direct impact ionisation requires a sufficiently high recollison energy for overcoming the second ionisation potential and is thus expected to be switched off for those trajectories with insufficient recollision energy.

8.3 Chapter 7: Strong Field Induced Dissociation

The experiments presented in this chapter investigated the molecular structure dependence of strong field induced dissociation in CO_2. In particular, the dissociative excitation of CO_2^+ was studied in the molecular frame as a function of probe laser intensity, ellipticity and polarisation with respect to the molecular bond. This allowed for the identification of the main excitation pathway consisting of tunneling ionisation from HOMO-2 followed by a parallel dipole transition from the second excited state B to the predissociating, third excited state C, whilst recollision excitation was shown to not play a role. Using laser induced impulsive alignment, the strong field induced coupling of the ionic states B and C could thus be controlled by the laser polarisation leading to a suppression of the fragmentation yield of up to 70 % with the laser polarisation perpendicular to the molecular axis. It was then found that the dissociation mechanism preferentially produces $CO^+_{(1,0)}$ ions with a very large branching ratio between $R_b = 0.8$ and $R_b = 0.9$. This implies that the molecular parent ion was mostly prepared in vibrationally excited states and remained in a linear configuration during the predissociation mechanism.

In the context of molecular physics, the identification of the dissociation pathway could be regarded as a minor discovery. For the strong field community, however, the observed coupling between the second and third excited ionic states represents an ideal playground for studying electronic and vibrational population dynamics in the strong field regime. This is an important ingredient for establishing spectroscopic techniques that tackle the ultrafast dynamics following the ionisation of complex molecules, for example. It should also be pointed out that the lack of recollision contribution is relevant in the context of using inelastic recollisions for inducing and controlling molecular dynamics. As was pointed out previously, recollision excitation may be controlled via the properties of the recollision event. The presented results show, however, that dissociation channels with relatively high appearance energy are needed to develop recollision controlled fragmentation schemes.

Future work in the context of the presented results could take two major directions. First of all they motivate the development of novel theoretical tools that describe electronic and vibrational population transfer in the strong field regime. Whilst particularly challenging due to the presence of the non-perturbative laser field, such tools are important for the development of strong field spectroscopy techniques as described in the previous paragraph. For this reason calculations lead by Olga Smirnova and

Misha Ivanov are currently underway modelling the obtained experimental results. This is expected to lead to a more refined understanding of molecular dynamics in the strong field regime and to test the interpretations of the dissociation mechanism presented in this chapter. Secondly, the employed experimental scheme could be extended to study fragmentation channels with higher appearance energy in order to access recollision induced channels. Here, one may look at fragments resulting from parent ions with higher charge states. However, this poses additional complications in the present experimental scheme due to the finite acceptance angle of the detector for collecting the fragments. This may be circumvented by installing a detector with a larger active diameter.

Reference

1. R.M. Lock, S. Ramakrishna, X. Zhou, H.C. Kapteyn, M.M. Murnane, T. Seideman, Extracting continuum electron dynamics from high harmonic emission from molecules. Phys. Rev. Lett. **108**, 133901 (2012)

MIX
Papier aus verantwortungsvollen Quellen
Paper from responsible sources
FSC® C105338

If you have any concerns about our products,
you can contact us on
ProductSafety@springernature.com

In case Publisher is established outside the EU,
the EU authorized representative is:
Springer Nature Customer Service Center GmbH
Europaplatz 3, 69115 Heidelberg, Germany

Printed by Libri Plureos GmbH
in Hamburg, Germany